The Iceberg Principles

The Truth About The Universe
And Your Place In It

Marni Spencer - Devlin

Table of Contents

Prologue

I was reborn when I discovered *the Iceberg Principles*—and I mean that literally.

I had been told I might have only a year to live. I had sold my sprawling, ocean-view home in Southern California, along with nearly everything I owned. What remained fit into two small suitcases. With those suitcases, I returned to Germany, the country of my birth, to make peace with my family before it was too late.

I sat at a small desk in my childhood bedroom, writing my memoir, *Crawling Into the Light*. It was my attempt to make sense of an existence that had felt more like a roller coaster ride than a life. I wanted it to be my legacy, the one thing that would remain after I was gone.

My life had been marked by rejection, molestation, and rape. When I was seventeen, I sought safety in the arms of a man who proved more awful than the world I was trying to flee. He lured me into heroin to control me. By the time I managed to escape, it was too late. I had become an addict. To survive, I turned to prostitution and crime. Eventually, I lost everything.

Hitting bottom shattered the last remaining illusion. With nothing left to protect, I turned myself in to the authorities, hoping for mercy. Instead, I was sentenced to prison.

Prison, unexpectedly, became my salvation.

It removed me from a trajectory that could not have been corrected from within. It stripped away distractions, momentum, and self-deception. When I was released, someone offered me a job in marketing, and I discovered an aptitude I hadn't known existed. Eventually, this led to the creation of a multimillion-dollar enterprise.

From the outside, it looked like a miraculous transformation. But inside, there was a black hole. The more success I achieved, the more hollow I felt. The chasm between how my life looked and how it felt widened until it became unbearable. At my lowest point, I seriously considered ending my life.

Then I became ill, and I wondered what it had all been for. Had there been a reason for my life, or was I just some cosmic joke? Had I really been in an accident, as my parents had told me over and over again? Had there been a purpose I had somehow missed? These questions felt enormous, and I wrestled with them as with a fiery dragon, sitting alone in that small room in Germany. I poured everything I had into my writing, terrified and trying to wrap my head around the fact that I might not be here by Christmas.

Then something happened that changed everything. It started innocuously enough. A friend who had stayed in touch with me, so I would not feel so alone, asked me a simple question: *"What is God?"* Not who, mind you, but what. I found the question intriguing. I sat down to formulate my answer, and suddenly words poured out of me.

It seemed my hand was writing faster than my mind could grasp the concepts. I had never thought about any of this before, yet the entire concept emerged fully formed, as if it had always been present. When I was done, I leaned back in my chair, stunned.

I immediately realized the implications of *the Iceberg Principles,* and they were profound. They fundamentally changed how I saw my life—and my death. Suddenly, my entire existence, all the craziness, and even all the pain made complete sense. Best of all, I knew I would never have to feel that way again. I knew beyond a shadow of doubt that no matter what would happen to me next, my life was not really ending.

Instead, it had just begun.

Introduction

L ife is hard—but it's better than the alternative!

There's a reason they call it the daily grind. If you want a decent life, you have no choice but to play along. Tired or not, you drag yourself out of bed and go to work. Few people get to do work they truly love. For most, it's the same routine, day after day. Somehow, even when you're doing everything right, you still worry. About money. About your health. About whether the relationship you're in will last—or whether you'll ever find the right one. You worry about the kids. Your parents. Most of all, you worry about whether you measure up.

You had dreams once. Most of us did. Somewhere along the way, you got tired, responsibilities piled up, and those dreams quietly slipped to the bottom of the list. Still, the feeling never quite goes away, does it? There's a sense— maybe the hope—that there was supposed to be something more than this.

Why, when we have so much, is it so hard to be human?

We live in an extraordinary universe. From the spiraling of galaxies to the rhythm of the seasons, structure is every- where we look. Expansion follows laws. Balance sustains systems. Cohesion holds complexity together. Nothing remains random for long; patterns reveal themselves at every scale. And yet, when it comes to our own lives, we often assume we are the exception. That we're abandoned in a vast universe and left to figure it out all on our own. We struggle as if there were no underlying design to draw on—no larger order to observe and learn from.

What if the universe is not merely a backdrop to human existence but a teacher? What if the same principles that allow stars to form, ecosystems to stabilize, and life to flourish could shed light on why we feel so lost, restless, or unfulfilled? This book begins with that simple premise:

The patterns we observe in the universe can teach us how to live with more clarity and ease.

Modern cosmology reveals something extraordinary about the universe we live in. Observable matter, including stars, planets, and everything we can see, makes up only about four percent of the universe's structure. The remaining 96% consists of dark matter and dark energy. It's called *dark* because we can neither see it, touch it, nor measure it. In fact, the only reason we know it exists is that it influences everything else. Galaxies rotate because of it. The universe expands because of it. Without this vast *invisibility*, the cosmos couldn't hold itself together. In other words, *what we can see exists only because of what we cannot.*

I propose that the same ratio also applies to human experience.

What we think of as *ourselves*—our body, personality, history, and circumstances—represents only a small fraction of who we are. It describes us, but it does not define us. Beneath the surface lies a deeper Blueprint that outlines the essence of who we are. This architecture defines our talents, passions, longings, and sense of purpose long before we appear in the visible world. What most profoundly represents us cannot be seen, yet it influences every aspect of our lives.

We are far more than what appears in the mirror. When we begin to see ourselves this way, something shifts. Fear loosens its grip. Curiosity replaces defensiveness. Even ordinary moments take on greater depth, not because life suddenly becomes easier, but because we stop resisting what is already happening. Through this lens, Consciousness is no longer an abstract idea. It becomes the medium through which life is lived.

This book explores that underlying structure. It isn't a scientific textbook, though it draws on cosmology, classical physics, quantum physics, biology, and neuroscience. It isn't a spiritual doctrine, though its insights echo across cultures and through time. And it isn't a memoir, though my life has certainly shaped these ideas.

It is a framework—a point of view that connects what we observe *out there* with what we experience *in here.*

Central to this framework is a simple premise: the visible Four Percent—the body, thoughts, roles, and circumstances—is real and essential, but it is not the Source. It is a mirror image. The Ninety-Six Percent is the fundamental, authentic architecture behind it all. When we are in alignment with that deeper structure, life tends to flow with clarity, vitality, and a quiet sense of rightness that doesn't depend on circumstances. When we are at odds with that part of us—through fear, conditioning, or inherited beliefs—we feel it immediately. Restlessness, anxiety, depression, and dissatisfaction are not signs that something is wrong with us; they are signals that something essential is being ignored.

This book isn't about fixing yourself. It's about understanding how reality works—and learning to live in coherence with it.

In the chapters that follow, we'll explore how unseen forces give rise to form, how Intelligence organizes matter, and why the same principles that govern galaxies also govern human lives. We'll examine why effort alone so often fails, why even great success can feel strangely empty, and why so many of us feel driven yet directionless.

You don't need to adopt a new belief system or abandon what you already hold dear. You need only stay open to the possibility that reality is far more coherent—and far more intimate—than you were taught to perceive.

The Iceberg Principles invite you to recognize the deeper design of who you already are.

Chapter One

The First Principle

The Substance Of The Universe Is Conscious Energy

When most people look up at the night sky, they see a black expanse dotted with twinkling lights. But have you ever considered that you are not just looking up at the stars? *You are gazing directly into eternity!* Those pinpoints of light are not decorations scattered across a ceiling. They are suns, some billions of years old, flung across distances so vast that human language breaks down long before it can describe them. The cosmos isn't just big. It is limitless in a way that dissolves every framework the mind relies on.

Astronomers estimate that there are at least 200 billion galaxies, each containing hundreds of billions of stars— many of which host planets, atmospheres, seasons, storms, and perhaps even life. When you gaze at the night sky, you see only the faintest dusting of what exists, the thinnest whisper of a universe so immense that if you picked any direction—up, down, or sideways—and somehow found a way to travel for a trillion years, you would still never reach the edge. You would simply keep going into more space, more stars, more galaxies, more unfolding mystery. There is no cosmic fence, no glass dome, no great wall with a posted sign that reads, *"End of Universe" (Prepare to Stop!)*.

But imagine if there were. What would lie beyond it? More space? Solid matter? Total emptiness? Another universe? The very question reveals the impossibility of a boundary. Infinity cannot be enclosed. *The idea of an edge exists only in the human mind,* which evolved within finite environments. The universe is not finite. It is not bounded. *It is not something that sits inside anything larger.* It is all there is, and everything we know, every breath we take, occurs within its endlessness.

When scientists speak of a multiverse, it often confuses more than it clarifies. For most people, imagining multiple universes is simply a matter of shifting the boundary—drawing a larger circle around the same infinite reality. Whether we call it one universe or many is essentially a matter of semantics. To existence itself, these distinctions are meaningless. Infinity is infinity. Eternity is eternity. The cosmos does not subdivide itself to accommodate the limits of human thought.

When we look into the night sky, we witness light that has traveled for millions of years to reach us. We see ancient history and the immediate present in the same moment. We gaze into distances so vast that our minds cannot grasp them, yet our hearts somehow can. The night sky is not a view; *it is a revelation.* A reminder that we live within infinity and that each of us is made of the same eternal light.

And once we understand this, even imperfectly, something shifts. The universe is no longer merely *out there,* a vast, impersonal expanse. It becomes both cosmic and intimate, majestic and familiar. It operates on scales so enormous—galaxies, black holes, star nurseries—that comprehension falters. Yet it also reveals itself in the smallest, most delicate details: the fractal symmetry of a leaf, the shimmer of a hummingbird's wing, the precise vibration of color, and the exact timing of a synchronicity that makes you stop and think, *"There's more going on here than meets the eye."*

Our universe is infinite, meaning it has no beginning or end. The human mind struggles with this idea. Maybe we can stretch our imagination far enough to picture something that never ends, but something that never started seems almost impossible. If it never began, how could it exist?

The difficulty isn't the concept itself—it's the tool we use to grasp it. The mind, marvelous as it is, operates within space and time. It evolved to track linear events: sunrise to sunset, birth to death, cause to effect. To us, everything has a before, a middle, and an after. Our stories rely on beginnings, middles, and ends. Our language relies on sequence. Our thinking relies on progression; one is followed by two, then three. Infinity breaks every rule the mind depends on. Outside of time, past, present, and future don't occur in a line—*they coexist.* Nothing is sequential. Nothing starts.

Nothing finishes. *Everything simply is, like two plus two is four.*

Trying to understand infinity with a linear mind is like trying to hold the ocean in your hands; the very act of grasping makes most of it slip away. And yet, even if we cannot fully comprehend infinity, we can sense it. We encounter it in the night sky, in mathematics, and in the strange fact that the farther we look into the cosmos, the more there seems to be. We encounter it in paradoxes, too: the universe expands yet has no center; it accelerates outward yet has no edge.

So, while we may question how and why, there's no denying the universe exists—because *we exist and are in it.* We are participants in the infinite, even if our minds can only glimpse it in fragments.

The Origins of the Universe

We don't know why the universe exists, but if there is truly nothing outside it—no external container, no larger framework, *no before* or *beyond*—then the universe must, in some sense, be its own Cause, Source, and Origin. It is the only thing that *is*, so whatever brought it into being must arise from within its own nature. This makes the universe not merely a physical entity but an active, self-generating process.

We also know that the universe is composed almost entirely of energy, with only a small fraction manifesting as matter. Even that matter is never still—it vibrates, pulses, transforms, and flickers in and out of quantum states. The entire cosmos is a dynamic, ever-evolving system governed by a set of exquisitely precise, interconnected physical laws —laws so consistent and elegantly tuned that life, chemistry, stars, and consciousness can arise from them. Scientists continue to uncover new layers of this order, revealing patterns that hold true across unimaginable distances and timescales.

When we look closely, we see that everything arises from the interplay of these laws across space and time. The universe behaves like a vast, self-regulating symphony, where each note is shaped by the one before it and, in turn, shapes the one that follows. Beneath what appears to be chaos lies a structure so precise that even the slightest variation in these fundamental constants would make life impossible.

What seems random at one level often reveals deeper patterns at another. Every event arises from causes and conditions, each woven into the next, forming a web of relationships spanning billions of years. A supernova in one corner of the galaxy produces heavy elements that drift through space, gather into new stars, form planets, and eventually become the calcium in our bones and the iron in our blood. The chain of causality is continuous and astonishingly coherent.

Some physicists argue that this apparent fine-tuning may result from selection effects or multiverse dynamics. Others see a deeper coherence in it. Perhaps the question remains open. However, for me, there's only one conclusion: *all of this cannot be accidental.* Order of this magnitude and elegance of this kind do not arise from random meaninglessness any more than the computer on your desk just popped up out of nowhere. There is an Intelligence at the foundation of existence—an underlying coherence that hints at purpose, intention, or, at the very least, Awareness. And where there is Intelligence, there must be Consciousness.

The Fundamentals of Consciousness

I believe Consciousness is fundamental. It did not suddenly emerge from matter as a lucky evolutionary accident. Rather, matter emerged *from Consciousness.* Consciousness is the substrate, the field, the unseen architecture underlying everything we perceive. It is not limited to brains or bodies; it is woven into the fabric of reality itself. It exists both impersonally and personally—impersonally as the laws and forces that steer the cosmos, and personally as the spark of Awareness within each of us.

Consciousness is the energetic spin that brings the atom into being, the organizing principle of star formation, and the quiet knowing at the center of every living being. It is both the canvas and the painter, the ocean and the wave. It is the foundation of existence, the silent Intelligence animating the universe, and the very essence of who and what we are.

Science cannot definitively explain how life began. It offers models, hypotheses, and elegant chemistry, but not certainty. The prevailing theory holds that in the early oceans, simple molecules gradually combined into more complex ones, eventually forming the first self-replicating structures—the precursors to DNA. Yet even this explanation leaves the central mystery unsolved. What caused matter, which had existed for billions of years without Awareness, to suddenly organize into living, sensing, self-perpetuating forms? How did chemistry become biology? How did biology become mind?

Modern neuroscience often holds that Consciousness is an emergent property—a byproduct of billions of neurons interacting in complex patterns. But this raises a deeper question: where did those billions of neurons come from, and why did they organize in such a way that Consciousness could arise at all? Emergence is a description, not an explanation. Saying that Consciousness *emerges* from neural activity is simply another way of noting that they occur together—it does not tell us how inert matter produces inner experience.

In fact, the more we study the brain, the more elusive the source of Consciousness becomes. Although regions of the brain correlate with different states of Awareness, no part has been identified as the generator of Consciousness. Neural firing explains electrical activity but not subjective experience: the feeling of a sunset, the ache of longing, the sense of self, the awareness of Awareness. How does electricity give rise to meaning? How does matter produce mind?

Some individuals retain Consciousness despite having only fragments of normal brain tissue, and some brain-dead patients report vivid, coherent awareness during near-death experiences—even when measurable brain function has ceased. These anomalies challenge the assumption that the brain creates Consciousness. Instead, they suggest that the brain may act more like a receiver or modulator—an interface through which Consciousness is expressed in physical form.

But science, which relies on measurable variables and reproducible data, is wary of anything it cannot quantify. Its mathematical framework describes only a narrow band of reality. Anything too subtle, nonlinear, or interconnected tends to fall outside the model. Yet even among the most rigorous researchers, one point of agreement remains: Consciousness is the most mysterious phenomenon in science. It defies the reductionist model at every turn.

Quantum physics has opened the door to this mystery. The discovery that a particle behaves differently when observed challenged the foundational assumption that reality exists independently of Awareness. The observer effect suggests that Consciousness is not a passive witness but an active participant in shaping physical outcomes. This does not mean the *mind controls matter* in a simplistic way, but it does mean that reality is not inert. It responds. It interacts. It changes under observation.

This is a profound discovery with far-reaching implications, yet we are still only scratching the surface of what Consciousness might truly be. Science may eventually develop tools to measure its effects more precisely, but for now, Consciousness remains a frontier—vast, enigmatic, and humbling. The fact that we can observe, question, and even wonder about its origin may itself be the greatest clue: that Consciousness is not a late arrival in the universe but its most fundamental feature.

The Spiritual View

Spiritual traditions, by contrast, focus not on equations or measurable phenomena but on the invisible forces that shape our inner lives. They explore meaning, purpose, intuition, awareness, and the nature of existence. Yet because they rely on interpretation rather than calculation, they are vulnerable to human subjectivity. Translations shift over centuries. Stories become symbols, then doctrines. Orally transmitted insights may splinter into countless variations. Even profound truths can be clouded by culture, power structures, or linguistic distortions. What begins as direct experience often becomes layered with metaphor, ritual, and opinion.

Despite these challenges, the spiritual impulse remains remarkably consistent across time and space. From ancient India to Greece, from the deserts of the Middle East to the forests of East Asia, humanity has always looked inward, compelled by the same enduring mystery: *What is Consciousness? Why do we have it? What is this Awareness that watches the world from behind our eyes?*

Each civilization, lineage, and spiritual tradition has sought to answer these questions in its own language and worldview. Their interpretations vary—some speak of the Soul, others of Spirit, Mind, Witness, Self, or pure Awareness. They disagree on origin stories and metaphysics, on the nature of God or the absence of God. Their descriptions are shaped by climate, culture, philosophy, and era. Yet when you place these teachings side by side, something remarkable becomes clear: they all point to the same central truth. They are like explorers mapping different sides of the same mountain, describing the terrain in different languages yet pointing toward the same summit.

For this book, the Hindu perspective offers perhaps the most expansive and comprehensive understanding of Consciousness as the ultimate reality. The ancient Upanishads teach that everything—matter, mind, time, space—arises from a single, formless, universal Consciousness called Brahman. This Consciousness is infinite, uncreated, and untouched by birth or death. It does not *exist within the universe*; rather, the universe exists *within it.* In this view, the personal self—Atman—is not separate from that cosmic vastness. The spiritual journey, then, is not about becoming something new but about remembering something ancient: that the consciousness within us is the same Consciousness that animates every being, every star, and every galaxy. Seen through this lens, the brain does not produce Consciousness any more than a lamp produces electricity; the brain is simply a temporary instrument through which the infinite expresses itself in form.

Christianity describes Consciousness using terms such as Soul, Spirit, Mind, and the image of God, yet, at its core, it shares a similar idea: human consciousness reflects the Divine. We are conscious because God—the ultimate Consciousness—is conscious, and we are made in His image. This means Consciousness is neither random nor accidental. It exists for relationship: with God, with others, and with creation. The Christian mystics understood this deeply. Meister Eckhart famously wrote, *"The eyes with which I see God are the same eyes with which God sees me,"* suggesting that Consciousness is a shared field, a meeting place where the human and the divine recognize one another through the same Awareness.

Islam offers one of the most beautiful images of Consciousness in all spiritual literature: God breathing His spirit into Adam. In that breath lies the divine spark within every person—an awakened Awareness of God's Presence, a sensitivity to what is true, and a living connection that guides and illuminates. In Sufism, this spark becomes the heart of a great love story: Consciousness as the Lover, yearning for the Beloved. The journey becomes one of remembrance. As the ego dissolves, what remains is the radiant Soul reflecting Divine Love. Islam teaches that Consciousness begins as Divine Breath and ends as Divine Return—a cycle of Awareness unfolding back into its Source.

The so-called *Heretics*—the Gnostics, esoteric Christians, and early mystics—offered a radical understanding that threatened established religious structures. They taught that Consciousness was a fragment of Divine Light hidden within each person, veiled by matter, suffering, and forget-fulness. In their view, the physical world was not false but incomplete—a dim reflection of a higher, more luminous Reality. For the Gnostics, the purpose of Consciousness was Awakening, not through doctrine or belief but through Gnosis: direct experiential knowledge of one's Divine Origin. While traditional Christianity emphasizes salvation through faith, the Gnostics emphasize salvation through remembering—remembering that the Soul did not originate here and that the spark in the chest is older than the stars. In their view, Consciousness does not need to be redeemed; it needs to be recognized.

When we place these spiritual traditions side by side, we see profound differences. Some claim Consciousness is eternal; others describe it as momentary. Some call it divine; others see it as lacking a fixed identity. Yet they all converge on one essential insight: Consciousness is the central mystery of human existence. It's the lens through which we experience the world and the key that opens realities beyond it. It's the part of us that feels most intimate and the part that feels most infinite. It's the one thing we cannot escape, yet cannot fully comprehend.

Understanding how the great traditions view Consciousness does not require us to adopt any single perspective. If anything, it expands our vocabulary for the most intimate experience we have: *Awareness*. It allows us to regard our own Consciousness not as a puzzle to solve but as a frontier to explore. Awareness is not merely something *we possess*; it is something *we* are. Whatever Consciousness ultimately turns out to be—energy, spirit, field, or fundamental property—it remains the most elegant, mysterious, and transformative force in the universe.

If Consciousness is fundamental—if it is not a byproduct of matter but the very ground from which matter, life, and awareness arise—then a deeper question naturally follows. Consciousness does not merely exist. It expresses, moves, and unfolds. Awareness, by its very nature, seeks to know itself, and the only way it can do so is through form.

This shifts how we understand the origin of the universe. The Big Bang can no longer be viewed as a random explosion of inert matter but as a moment of articulation—Consciousness differentiating so it could experience itself. Not an accident, but an expression. Not chaos, but emergence with direction. Seen this way, the universe is not something that happened *to Consciousness*. It is something that happened *because of it*.

To understand reality, we must move beyond asking what Consciousness is and begin asking why it takes form at all. That inquiry brings us to the second principle.

Chapter Two

The Second Principle

The Big Bang Was the Result of Consciousness Seeking Expression

If Consciousness is fundamental, then differentiation into form may be understood as a way for Awareness to encounter itself through contrast. When we look beneath every scientific theory, every spiritual teaching, and even every personal experience, a single unifying idea begins to emerge: the primary focus of Consciousness, whether in the universal or personal realm, is to *become more conscious of itself.* Everything Consciousness does aims toward that end. It seeks to explore, create, imagine, dream, philosophize, rewrite reality, or wander into its own mysteries through contrast—all in an effort to understand itself more deeply.

This is not mere poetry. It is a pattern so consistent that it appears across evolution, psychology, cosmology, spirituality, and the unfolding of every human life. Consciousness, whatever its origin, behaves as if its deepest instinct is self-revelation. Its curiosity, creativity, questions, and break-throughs all point toward the same driving impulse: *the desire to know itself.* To see this clearly, we need only observe what Consciousness actually *does*.

Consciousness is driven to explore. From the earliest single-celled organisms moving toward light to humans strapping themselves into rockets and launching into the unknown, Consciousness reveals a persistent drive to push beyond its own boundaries. It seeks to discover what lies beyond the familiar, what is possible, and what exists beyond the horizon. This urge cannot be reduced to survival alone. Humans explore even when there is no biological advantage, when the adventure is dangerous, costly, or purely imaginative. We map oceans, decode DNA, and peer into black holes for one reason: Consciousness is drawn to the unknown because the unknown is a mirror. By facing what it does not yet comprehend, Consciousness expands. *Exploration is expansion.*

Consciousness seeks to create. Creation is not an accessory to life but a fundamental expression of Awareness. When an artist paints, a scientist formulates a theory, a child invents a game, or a culture constructs a myth, *Consciousness externalizes the internal—making visible what was once invisible.* Every work of art, every invention, every story ever written is Consciousness studying itself from a new angle. Creation is not separate from self-understanding; it is self-understanding taking form. Through creation, Consciousness discovers what it contains.

23

Consciousness dreams. Neuroscience describes dreams as ways to reorganize memory, regulate emotion, and rehearse potential futures. But something deeper unfolds in that nocturnal theater. In dreams, Consciousness reshapes reality, explores symbolic landscapes, tests new identities, and reveals hidden layers of its own psyche. It becomes both the storyteller and the protagonist, the architect and the inhabitant of the dream. In this sense, dreaming is Consciousness learning through imagination—*encountering itself in symbol and metaphor.*

Consciousness is compelled to philosophize. Humans question everything—not because life demands it, but because Consciousness is wired for reflection. We ask: *Who am I? Why am I here? What is real? What does it mean to exist?* Philosophy is Consciousness turning inward, studying itself with the precision of a scientist in an experiment. It is the mind attempting to map the mind. Even when answers remain elusive, the questioning itself deepens Awareness. The act of asking becomes a catalyst for growth.

Consciousness also seeks to reshape reality. The human mind does not merely observe the world—it transforms it. We alter landscapes, build societies, translate experiences into stories, break inherited patterns, heal trauma, and imagine futures that never existed before. Consciousness is not content to accept reality as presented; it continually reimagines and redesigns it. Every act of personal or collective transformation is Consciousness declaring, *I am more than what I appear to be.* Rewriting reality is a form of self-evolution. By reshaping its world, Consciousness experiments with reshaping itself.

Ultimately, Consciousness is irresistibly drawn to the one territory that remains forever mysterious: *itself.* This is why humans meditate, pray, journal, seek therapy, ask existential questions, or enter altered states. Consciousness turns inward, attempting to perceive the perceiver, witness the witness, and locate the *"I"* behind the *"I Am."* It asks: What is this Awareness that notices everything? What is the source of my thoughts? What remains when all roles and identities fall away? Each spiritual practice, each reflective inquiry, and each inner search is Consciousness attempting to observe itself directly.

When we step back, the pattern becomes unmistakable. Consciousness explores the outer world to understand its capabilities. It creates to witness its own imagination. It dreams to chart its inner terrain. It philosophizes to inquire into its nature. It reshapes reality to test its boundaries. And it turns inward because it senses that something within it is deeper than thought, narrative, identity, and even the world itself.

This is not random behavior. It is the signature of a force that evolves by understanding its own depth. If Consciousness is always striving to become more self-aware, then human life—our creativity, suffering, breakthroughs, and desires—is part of a much larger unfolding. We are not simply living individual lives; we are participating in the universe's project of self-discovery. *Through us, the cosmos observes itself. Through us, Awareness recognizes its own reflection.*

Every insight, every act of courage, and every moment of Awakening contributes to the expansion of Consciousness as a whole. We are not separate from this evolution: we are the evolution. *Consciousness expands through experience, deepens through inquiry, and flowers through creation.* Every step toward greater understanding is Consciousness fulfilling its most ancient aim—the desire to know itself more completely.

The Big Bang

From the standpoint of modern science, the story of the universe begins with an explosion. A single point of unimaginable density suddenly expanded, giving rise to space, time, matter, and the laws that govern them. This moment—the Big Bang—has become so familiar in our cultural vocabulary that we rarely pause to consider what it implies. We speak of it casually, as though a universe simply appearing from nowhere is something we should accept without question. Yet when we look closely, we discover that the Big Bang is not a complete explanation. It is a doorway. It tells us *what happened,* but not *why.* It offers a description, not a cause. Science can describe the unfolding, but it cannot explain what set that unfolding in motion.

If we trace the chain of events backward, through the formation of galaxies, the ignition of stars, the cooling of planets, the emergence of complex chemistry, and the appearance of life, we eventually reach a boundary where sci-ence must stand still. Beyond that boundary lies a silence, a condition without space or time, without distance or dura-tion, without form or energy. We *cannot call it before,* be-cause time had not yet existed. We *cannot call it something,* because no *thing* had yet appeared. Yet, this absence of physical qualities does not mean the absence of all qualities. On the contrary, whatever preceded the Big Bang must have contained, *in a latent form, the potential for everything that followed.* It must have held the seeds of geometry, sym-metry, lawfulness, creativity, and relationship. It must have possessed the inherent capacity to become a universe.

For centuries, philosophers and mystics have argued that the most fundamental ingredient of reality is not matter, energy, space, or time—but Consciousness. Not Conscious-ness as thought or emotion, but *as the raw capacity for Awareness,* the field in which experience becomes possible. In this view, Consciousness does not arise from the uni-verse; the universe arises within Consciousness. Matter does not create awareness; awareness gives rise to matter. Con-sciousness is not a late arrival on the cosmic stage. It is the stage itself.

Imagine an infinite stillness without form or boundary. Nothing in motion, nothing to compare, nothing to measure. A pure, undivided Awareness so complete in itself that *it cannot yet perceive itself.* It is like a still lake with a perfect-ly smooth surface. The lake exists, but nothing disturbs it enough to reveal its depth. It cannot know itself because there is nothing to reflect it. For Awareness to become self-aware, something must appear *other than itself.* A contrast must emerge. A surface must ripple. A mirror must be created. Without distinction, there is no recognition. Without relationship, there is no perspective. Without movement, there is no experience.

If Consciousness is the substrate of existence, then the birth of the universe can be seen as a natural expression of its desire to know itself more fully. The Big Bang ceases to be a random accident and becomes the moment when Awareness created the conditions for reflection. With that first expansion, time came into being, providing a sequence in which events could unfold. Space arose, providing dimension and separation. Forces and fields took shape, creating the possibility for form and complexity. The laws of physics crystallized, allowing relationships to develop across vast distances. In a single gesture, *Consciousness produced the entire framework* that would eventually allow it to look back on itself.

This idea gains strength when we examine the scientific picture in greater detail. Moments after the Big Bang, the universe was governed by perfect symmetry—a state so perfectly uniform that nothing distinguishable yet existed. But symmetry alone carries no meaning. A completely uniform field cannot give rise to structure. It is the breaking of symmetry—minute fluctuations in the early universe—that allowed matter to clump together, stars to ignite, and galaxies to form. These fluctuations were so precise that the universe balanced on a razor's edge between collapse and dispersion.

If they had been even slightly different, stable structures would not have formed. The physical laws themselves—gravity, electromagnetism, the strong and weak nuclear forces—are so finely tuned that the universe seems predisposed toward complexity, as though it had a tendency to organize itself, build upon itself, and become increasingly capable of supporting Awareness.

Over time, particles combined into atoms, atoms into molecules, and molecules into living cells. Life emerged not as an anomaly but as an extension of the universe's inherent drive toward complexity. Once life appeared, it began to sense, respond, adapt, and evolve, and Consciousness became ever more self-aware. Nervous systems developed. Brains formed. Eventually, Consciousness, the silent field behind everything, acquired an instrument—a human mind —through which it could consciously observe its own creation. When a human being looks at the night sky, it is Consciousness examining the body in which it resides. When we ask why the universe exists, it is Consciousness posing the most intimate question it can: *"What am I?"*

We fulfill our purpose simply by existing conscious.

Some interpret this perspective metaphorically. Others take it literally. Yet the logic remains elegant. Everywhere we look, the universe reveals a pattern: it unfolds toward Awareness. Hydrogen becomes stars. Stars become supernovae. Supernovae create heavy elements. Heavy elements form planets. Planets host life. Life evolves brains. Brains generate awareness. Awareness asks questions. The Cosmos becomes conscious of itself through the beings it produces. This does not diminish the scientific story; it enriches it. It suggests that Consciousness is not an afterthought but a thread woven through the entire fabric of existence.

Despite monumental advances, modern neuroscience still cannot answer the central question of Consciousness: how does physical matter produce subjective experience? How do measurable electrical signals create the feeling of a sunrise, the sound of music, the tenderness of love, the ache of longing? The most honest answer science can offer is that it doesn't know. Correlations exist—certain brain states accompany certain experiences—but correlation is not creation. The brain may be a receiver, a translator, a filter, or a lens through which Consciousness is expressed, but it does not explain Consciousness itself.

Quantum physics adds another layer to this picture. At the smallest scales, particles are described not as fixed objects but as probabilities that resolve into measurable outcomes through interaction. What becomes real depends, in part, on how a system is engaged. Reality is not entirely passive; it participates in relationship. This does not mean that human thought controls matter or that Consciousness overrides physical law. It suggests something more subtle: that observer and observed are not completely separate, and that the act of interaction shapes how events unfold. The universe, at its foundations, appears less like a rigid machine and more like a dynamic process in which potential becomes form through connection.

Seen through this expanded lens, the Big Bang becomes the opening movement of a long symphony, one whose theme is the evolution of Awareness. The early universe was simple, nearly featureless. Over billions of years, complexity increased, Awareness deepened, and Consciousness found more intricate and expressive forms. Human beings are among those forms—not the final expression, not the pinnacle, but a current stage in the universe's unfolding capacity to reflect itself.

If this is true, our presence here is not incidental. We are expressions of the same Consciousness that gave rise to galaxies. We are made of the same ingredients that formed the stars. The atoms in our bodies were forged in the hearts of ancient suns. The capacity to think, imagine, create, and question is not a biological quirk; it is the universe continuing its search for self-understanding through us.

The implications are both humbling and empowering. Humbling because we realize we are participants in something vast, ancient, and intricate beyond comprehension. Empowering because we carry within us the same essence that shaped the cosmos. Consciousness did not arrive in the middle of the story—it authored it. And the same force that set the universe in motion is the force that thinks through us, dreams through us, loves through us, and wonders through us.

The more deeply we consider this, the more natural it becomes to view the universe not as a cold, indifferent expanse but as a living process, animated by a fundamental

Intelligence. That Intelligence expresses itself impersonally as physical law and personally as qualities we associate with human life: curiosity, creativity, imagination, and the desire for meaning. When we experience awe, inspiration, or a sudden insight, we are touching the same Source that flared into existence 13.8 billion years ago.

Consciousness evolves by becoming more self-aware. The universe expands not only in size but also in understanding whenever a mind gazes into its depths. Every breakthrough, every question, every act of curiosity is Consciousness pushing its frontier forward. Through science, art, philosophy, and spirituality, Consciousness examines itself from different angles. Through each human life, it gains a unique lens through which to see. No two people are identical because each represents a distinct expression of this unfolding awareness. It never repeats itself, except where repetition is the point itself. What would be the point otherwise?

This naturally leads to the conclusion that our individual purpose is not separate from the universe's purpose. If Consciousness created the universe to know itself, then each of us inherits that same drive. We are born with a longing to understand ourselves, to express who we are, and to translate invisible potential into visible form. When we feel restless, incomplete, or unfulfilled, it is often because we are living only from the surface layer of who we are. We sense a yearning deep within but have not yet learned how to access it. When we express our gifts, contribute to others, create beauty, or pursue knowledge, we feel aligned because our inner and outer worlds move into coherence.

Chapter Three will explore this idea through the lens of the cosmic ratio—the ninety-six percent that remains unseen and the four percent that is visible. Just as the universe is shaped by forces that cannot be directly observed, so are we. Our physical form is only a small fraction of our total identity. The vast, unseen portion holds our potential, intuition, creativity, emotional intelligence, purpose, and deepest truths. Understanding this ratio allows us to navigate our lives differently. When we recognize that the surface is merely a reflection, we learn to work with the immense reservoir beneath it.

But before we can understand how this ratio applies to human life, we must recognize its presence in the universe itself. The Big Bang was not merely the beginning of space and time. It was the moment Consciousness created a mirror to know itself. Everything that followed—every star born, every life form that emerged, every new idea that passes through a human mind—is part of that same unfolding. When we contemplate the origin of the universe, we are contemplating the origin of ourselves. The story of the cosmos is the story of Consciousness waking up. In Chapter Three, we will turn to the structure that makes this awakening possible: the delicate, astonishing balance between what is visible and what is not—the cosmic ratio that shapes galaxies, governs nature, and lives within every human being.

Chapter Three

The Third Principle

The Cosmic Ratio Is the 96:4 Percent Relationship Between the Quantum and the Physical World

When astronomers first began weighing and measuring the universe, they expected a certain kind of simplicity. If you add up all the stars, galaxies, gas clouds, dust, and other visible matter, you might imagine that accounts for everything that exists. But the more carefully scientists looked, the stranger the picture became. Galaxies were spinning too fast. Clusters of galaxies were held together by more gravity than visible matter could supply. The universe was expanding not just steadily but at an accelerating rate, pushed by an immense unknown force. Bit by bit, the numbers forced an uncomfortable conclusion: *almost everything that exists is invisible.*

According to our best current models, only a small fraction of the cosmos—roughly four percent—is ordinary matter, the kind of stuff that forms stars, planets, bodies, and trees. The remaining ninety-six percent consists of something we cannot see directly: dark matter, which holds galaxies together, and dark energy, which seems to drive the expansion of space itself. We don't know exactly what these things are. *We only know that without them, the universe we observe could not exist in its present form.*

This proportion—four percent visible, ninety-six percent invisible—is known as the Cosmic Ratio. It is more than a technical detail in cosmology. It points to a fundamental feature of the structure of reality. The universe is not built around what we can see; it is built around what we cannot see. The visible is the exception, not the rule. The manifest is the surface, not the depth. The measurable is only the tip of the iceberg, not the iceberg itself.

When you first hear this, it might sound abstract, even remote from daily life. Who really cares how much of the universe is dark matter versus ordinary matter when the main concern is getting through the day? But the Cosmic Ratio matters very much because it is not just *out there* in space; it is mirrored in us. The same asymmetry between the visible and the invisible, the surface and the depth, appears again in human beings. Until we grasp that, we won't understand why life feels the way it does. *Why do we experience so much frustration and hunger, even when it looks like we have more than enough?*

Before we bring the Cosmic Ratio down to the human scale, it is worth staying with the universe for a moment longer. The more we discover, the clearer it becomes that reality is woven from layers we cannot perceive directly. Atoms, once thought to be solid little marbles, turned out to be mostly empty space, with tiny particles whirling in complex fields. Those particles, in turn, behave more like waves of probability than little billiard balls. Beneath the apparent solidity of matter lies a world of energy, information, and interaction that reveals itself only under the most careful investigation.

And yet, despite the complexity and mystery, the universe holds together. Galaxies do not fly apart. Stars do not randomly wink in and out of existence. Physical laws remain stable over billions of years. There is a coherence to the whole that cannot be accidental. An infinite, self-sustaining system cannot harbor flaws that compound endlessly; if it did, the entire structure would unravel. Whatever this universe is, it is clearly built on precise principles that preserve its integrity.

That does not mean everything is pleasant or tidy. Supernovas explode. Black holes devour matter. Planets collide. The universe is not gentle. But it is consistent. Its laws do not contradict themselves. It does not casually self-destruct. There is a deep reliability and an underlying Intelligence at work.

Many traditions have used the word *God* to name this underlying Intelligence, but for countless people, that word has become burdened with images of a person-like figure in the sky prone to wrath, erratic moods, and unpredictable allegiances. It has been used to justify wars, prejudice, and fear. For the purposes of this book, I will not rely on that term, not because I dismiss what it points to, but because I want to disentangle the reality we are exploring from the dogmas that have obscured it. I am not interested in arguing for or against any religion. I am interested in the nature of the universe we actually live in and the nature of the Consciousness that animates us.

We know, for a fact, that we are conscious. We are aware, and we are aware that we are aware. We also know that we exist in a universe governed by consistent laws, organized into patterns, and that these patterns give rise to life and mind. We can debate their origin, but we cannot deny their presence. There is a driving force behind it all, a tendency toward form, complexity, and relationship. There is something that keeps electrons in motion, hearts beating, ecosystems adapting, and galaxies swirling. The universe is not static. It is in constant motion and constant exchange. Even what we call *empty space is not empty.* It is a cohesive field that seethes with quantum fluctuations we barely understand.

The sciences have done a remarkable job of describing what this motion looks like. They map forces, measure particles, and simulate the evolution of the cosmos. But there is one question they have never fully answered: *What is the phenomenon called life?* Not biologically, but fundamentally. What is this animating presence that breathes through the cells of a body, that gives an organism not just structure but sentience, not just chemistry but experience?

At some point, we must acknowledge that this question touches something beyond the reach of instruments. Consciousness is not a thing we can point to on a scan. It is the condition that makes pointing, scanning, and knowing possible. It is the field in which thought arises. It is the Awareness that notices both the outer and the inner worlds. When you reflect on your life, you realize that everything has changed—your body, your preferences, your roles, your beliefs—but something constant has been present through it all: a quiet Presence —the *I Am* behind the scenes. That *I Am* is Consciousness.

Consciousness, by its nature, is not measurable. It cannot be weighed or located in space. And yet it clearly exists; it is the most intimate fact we know. If even one thing is non-measurable, the assumption that only the measurable is real is obviously incomplete. The non-measurable must be part of the universe's fabric as well. In studying the Four Percent—the visible, dimensional realm—it is therefore foolish to ignore the Ninety-Six Percent simply because we lack the tools to quantify it.

The Cosmic Ratio, then, is not just about dark matter and dark energy. It is about the divide between what the intellect can comfortably grasp and what lies beyond its reach. The intellect is an extraordinary tool. It helps us navigate linear time, compare options, solve problems, and coordinate actions. It is like a car's instrument panel: it tells us how fast we're going, how much fuel we have, and whether the engine is overheating. But no one would confuse the instrument panel with the driver, the journey, or the landscape. In the same way, our thinking mind is not the Source of life. It is the interface.

Because the intellect processes only data from the Four Percent world, it tends to deny or distort anything that does not fit its categories. When confronted with the non-dimensional—the infinite, the eternal, the interior—it balks. Infinity offends its sense of sequence. Eternity contradicts its understanding of time. The idea that Consciousness might precede matter rather than arise from it feels backward to a mind trained to start with what it can touch.

But if we look honestly at our experience, we see that form is always preceded by the intangible. Everything we create begins as an idea. A painting, a business, a relationship, a book—all start as a pattern in Consciousness before becoming visible. An idea is never just an electrical impulse in the brain. Instead, it appears as a complete blueprint: a melody, a sentence, an image, a solution that feels as if it dropped into awareness from beyond ordinary thought. Where did it come from? We can call it imagination or inspiration, but what inspired it? We can call it intuition, but where did it come from? Whatever we call it, it is clear that something new entered the mind from a source that cannot be reduced to *the rearrangement of old data.*

This is where the Ninety-Six Percent becomes personal. The intangible universe *is not only out there* in the form of dark matter and an unknown field; *it is also in here*, as the vast interior territory from which insights, impulses, desires, and potentials arise. In later chapters, we will explore this invisible side more directly, but for now it is enough to recognize that the Four Percent world we can see—bodies, biographies, circumstances—is only a small part of the story. The bulk of who we are, and the bulk of what shapes our lives, is intangible and lies beneath the visible surface.

The Cosmic Ratio, then, is a structural clue. It tells us that any attempt to understand life by looking only at what is visible will always be vastly incomplete. The visible depends on the invisible. In fact, it is its reflection, its mirror image. The expressed originates in the unexpressed. The actual arises from the potential. Just as galaxies are shaped by forces we cannot see, human lives are shaped by dimensions of being we cannot reduce to mere physical descriptions.

There is another implication of the ratio that is easy to miss yet critically important. In a universe where so much of reality is hidden, certainty becomes dangerous. When we believe our current understanding encompasses the whole, we close ourselves off to the Ninety-Six Percent that does not fit. This is as true in science as in spirituality and psychology. Dogma—the insistence that our partial view is complete—is the enemy of truth.

If we accept the Cosmic Ratio as a working premise, humility becomes the only reasonable stance. We can say, *"This is what we know about the Four Percent,"* and perhaps even, *"These are some of the ways the Ninety-Six percent seems to show itself,"* but we stop pretending we have all the answers. Instead, we become collaborators with a universe still revealing itself. We listen, experiment, and re-fine. We allow for mystery without collapsing into supersti-tion, and for science without disintegrating into reduction-ism.

So where do we, as individuals, fit into all this? In many ways, our existence mirrors the structure of the cosmos. The part of us that is visible is small: our bodies, our roles, our histories, our public personas. This is the Four Percent. It is important, but it is not the whole by any means. The larger part of us—our inner world, our unexpressed gifts, our po-tential, our deep knowing, *the Signature Blueprint of who we are*—is invisible. It is the Ninety-Six Percent. We feel it as a kind of background hum, a sense that there is more to us than others see. That there is more to life than the rou-tines we perform, and that desires and yearnings are smol-dering beneath.

When the expression of the Four Percent doesn't reflect the truth of the Ninety-Six Percent, life begins to hurt. We feel restless, out of place, and unsatisfied. I was at the height of my success, making more money than I ever dreamed of, surrounded by all the finest things in life. And yet I was completely unhappy, to the point of contemplating suicide. I chided myself for being so ungrateful, even though I had it all. I hated myself for acting like a spoiled child. What I didn't realize until much later was this: All the money and business success meant nothing because they didn't align with who I really am. We may have achievements, relation-ships, and a functioning daily life, yet something within us knows we aren't living up to our full potential. We are acting out a partial script. The frustration we feel isn't a personal failure; *it's a signal that the visible and invisible parts of us aren't matching up.*

The Cosmic Ratio helps us understand this pain in a new way. Instead of seeing it as a sign that we are broken, we can see it as evidence of how much greater we are than we imagined. We are Four Percent beings carrying a Ninety-Six Percent destiny. No wonder it feels cramped.

In this chapter, we have stayed largely at the level of structure. We have examined how the universe is composed and how that composition points to an imbalance between what can be observed and what cannot. We have suggested that Consciousness is not a late arrival in an otherwise dead universe but a fundamental aspect of reality and its very reason for being. We have noticed that everything we create, from a painting to a life, begins in the invisible, grounded in an underlying structural design, before it becomes visible. We have suggested that human dissatisfaction often stems from a mismatch between our deeper Blueprint and our surface expression. The sense we often have that there should be more comes from living lives too small for our larger destiny.

In the next chapter, we will bring the Cosmic Ratio into everyday life. We will look closely at the Four Percent—the visible self: the body, the personality, and the story we tell about who we are. We will examine why this Four Percent identity has come to dominate our sense of self, how it shapes our choices, and why it is inherently unable to satisfy us on its own. Only when we understand the limits of the Four Percent can we appreciate the necessity of the Ninety-Six. Only when we see how small the surface really is can we begin to turn toward the depth that has been there all along.

Chapter Four

The Fourth Principle

The World We Inhabit Is Only Four Percent of Reality

With the Big Bang, time, space, and the physical universe came into existence. From the perspective explored in this book, this was not merely an explosion of matter but the emergence of a mirror—a means by which Consciousness could express itself and perceive its own reflection. The physical world was never meant to be the ultimate reality. It was meant to be an interface. As it stands, we live our entire lives inside this interface and take it as solid, reliable, and unquestionably what is *out there*. Yet, when we examine it more closely—scientifically, psychologically, and experientially—we discover something astonishing: the world we live in does not exist the way we think it does at all.

A Projection of Reality

Earlier, we said that the universe is mostly energy, with a small fraction appearing as matter. Modern physics has revealed that even what we think of as solid matter is far less solid than it seems. At the subatomic level, particles are not tiny billiard balls but excitations of underlying quantum fields. An atom, the building block of everything in the physical world, is mostly empty space structured by probabilistic relationships.

Werner Heisenberg once remarked that *atoms are not things* in the classical sense but are better described as patterns of possibility. In quantum mechanics, an electron is represented by a wave function—a mathematical distribution that describes the probabilities of where it may be detected upon measurement. Before it is observed, an electron does not occupy a definite position. Instead, it exists *in a superposition* of infinitely many possible states. Only upon observation does it collapse into a single, definite position.

It is important to clarify what *observation means* in this context. In physics, observation does not require a conscious mind. It refers to physical interaction—when a quantum system interacts with another system, such as a measuring device or its surrounding environment. These interactions lead to what physicists call decoherence, which explains how mere *probabilities* could give rise to the stable, classical world we experience. This means that, thankfully, the macroscopic world does not depend on our thoughts to remain intact. Its stability emerges from constant physical interactions occurring at scales far beyond our awareness. So, when we look at the world around us, we are not creating it with our attention. The external world exists independently of any individual observer. However, our experience of that world is constructed.

While physics describes how matter behaves, neuroscience explains how our experience arises from sensory input and neural processing. What we call *reality* in daily life is an interpretation constructed by the brain.

The brain is not a passive receiver of information, faithfully recording what is *out there.* Increasingly, neuroscience understands perception as *predictive.* The brain functions as an inference engine. It continuously generates models of what it *expects* to encounter and compares incoming sensory signals with those predictions. When the data roughly match expectations, the brain fills in the gaps seamlessly. When they do not, the model updates—but usually within surprisingly narrow limits.

In this sense, *our perception is less like a camera and more like a controlled hallucination constrained by sensory input.* The world we experience is literally assembled by the brain. Stability is not proof that its perception is perfect. It is evidence of its efficiency.

Vision offers a striking example. Only a small central region of the retina—the fovea—provides sharp, high-resolution detail. Most of the visual field is low resolution. Yet when we look around, we experience a seamless, detailed panorama. This is because the brain actively fills in missing information using context, memory, and expectations. Rather than processing every pixel in real time, the brain constructs an *adaptive approximation* that updates many times per second. This strategy conserves enormous metabolic energy while preserving functional accuracy.

The same principle applies to color and sound. Color does not exist in the external world as *red* or *blue*. Light has measurable wavelengths, but hue is a perceptual experience generated deep within the brain only when those wavelengths are processed by our neural circuits. Two people may agree they are looking at a red flower, but each person's experience of what *red* means is entirely subjective.

Sound works similarly. Outside the body, pressure waves travel through the air. The universe itself is entirely silent, but when those waves interact with the eardrum and are interpreted by the brain, we experience sound as music or spoken word.

Philosophers call these subjective qualities qualia—the felt textures of experience: the redness of red, the sweetness of fruit, the timbre of a violin. These are not properties, actual truths floating in space. They are experiences generated only by the interaction between the external world and the nervous system. This does not mean the world is unreal. There is an objective reality independent of us. What varies is how the experience is rendered within us.

The Four-Percent world—the visible, measurable surface of experience—is not what it seems. It depends entirely on our personal rendering. It is evolutionarily refined, deeply immersive, and extraordinarily useful. Yet it is not the full story of what exists. It is merely the interface through which a far richer reality is interpreted and lived.

Think of it like watching a movie. We respond emotionally as if events were truly happening, even though they are not. We feel fear, relief, love, and grief. Our heart rate changes. Our muscles tense. We cry real tears. And yet nothing is actually happening on the screen—only patterns of light and shadow. The experience is generated entirely within us. The Four-Percent world operates in much the same way. It is compelling, convincing, and emotionally powerful, yet it remains a *projection* interpreted by the mind.

The Inner Narrative

This becomes especially important when we turn our attention inward. Pause for a moment and consider a simple question: *where do your thoughts come from?* From the moment you wake up, the inner narrative begins, but do you decide what you are going to think next, or do thoughts simply appear in your mind? The constant commentary evaluates, plans, judges, worries, and rehearses for you. It is always there, so you never question this voice. It's inside your own head, so you assume it is who you are. I mean, who else should it be! It speaks continuously and with such great urgency that its authority feels self-evident. But here's the truth: The constant chatter is simply an output of the brain, generated purely on autopilot.

Shaped by past experiences, emotional memories, and conditioning, it is not even a response to the present moment. Rather, it is a recycled interpretation of the past, replayed in real time. The brain is extraordinarily efficient, and one way it conserves energy is by reusing familiar patterns. It would rather repeat an old thought than generate a new one. As a result, much of what you think today is yesterday's thinking carried forward. In other words, it is old news.

Here is the problem: when a thought arises, you identify with it, *"This is me."* Suddenly, awareness collapses into content. Random, fearful thoughts solidify into an identity; you think of yourself as an anxious person. A self-critical thought becomes a global judgment about who you are. A momentary doubt hardens into a perceived limitation that follows you forever. This fusion of awareness and thought is one of the most powerful illusions, one that literally imprisons us in the Four-Percent world.

There is also a second source of thought that often goes unnoticed: the collective mental environment. Just as the body breathes the air around it, the mind absorbs the emotional and conceptual atmosphere of culture, media, and community. Anxiety, outrage, scarcity, comparison, and fear circulate continually. These thoughts arrive without labels; they simply appear. Because they are inside your head, they feel familiar, and you assume they are your own. The mind then personalizes them. The general fear about the economy becomes: *"What if I lose my job?"* Collective fear becomes a private catastrophe. All these thoughts, which were never your own, then shape your perceptions and decision-making. They determine your behavior without your awareness.

At the neurological level, this process is reinforced by the reticular activating system in the brain. This network of neurons in the brainstem serves as a vital gatekeeper of attention. At any given moment, the nervous system is bombarded with billions of bits of data, far more than it could ever consciously process. The reticular activating system determines what is allowed through and what must be filtered out. In other words, it decides what should matter to you long before you have a chance to weigh in. The criteria it uses to make these fundamental decisions are impersonal: relevance to survival, emotional charge, and familiarity. Whatever aligns with your existing beliefs, concerns, and unresolved fears is prioritized. Whatever does not is relegated to the back burner or ignored entirely. This is not a flaw; it is necessary. It is how the brain maintains coherence. The only problem is that this filter operates on unconscious, inherited assumptions, which determine the only information your mind presents to you.

Once an interpretation of reality has been adopted—*the world is unsafe, opportunities are scarce, I must be vigilant*—the nervous system begins scanning relentlessly for confirmation of these assumptions. Even neutral events are filtered through this lens; any ambiguity is resolved in favor of potential danger. Just in case. Any information that contradicts this narrative is filtered out before it reaches your awareness. Over time, the output that does reach you feels obvious. *"Yep, ya can't trust anybody!"* This then no longer appears as perspective but as reality itself. This is how collective anxiety becomes your personal fate.

At the center of this process sits the amygdala, an ancient brain structure that evolved to keep us alive. It responds far faster than conscious thought and cannot distinguish between physical danger and psychological threat. A predator in tall grass and an uncomfortable email trigger the same alarms. The body reacts before the mind can reason. This is why one critical comment outweighs ten compliments, why vague worry eclipses concrete safety, and why the ego, the intellect identifying with itself, becomes preoccupied with risk, comparison, and control. Its task is preservation. Growth feels dangerous. Expression feels risky. Change threatens stability. Even joy can feel unsafe if it expands us beyond the familiar.

The Terrible Conflict of the Four Percent

This, in a nutshell, creates the central tension of our lives. The Four Percent orientation is built to seek safety, familiarity, and control in a seemingly dangerous world. But there is something within us all that inexorably seeks expression and expansion. Something that yearns for fulfillment. Indeed, the Buddha identified unfulfilled *desire* as the root of our greatest suffering. This burning desire to express who we are can perhaps be suppressed, but it can never be silenced because it is the larger part of us. For most people, this conflict manifests as unexplainable suffering. Sometimes it hurts even more when, to the naked eye, everything looks just fine, yet it still feels as if something is missing. There is a strange tug-of-war between a general sense of overwhelm and the feeling that there should be more to life. Depression, anxiety, boredom, addictions, and every kind of self-sabotage are many types of emergency brakes applied the moment growth approaches.

The ego is not the enemy; it is merely a tool—an *interface designed for survival, not destiny*. The problem arises when the ego mistakenly identifies with the whole person. The Four-Percent world was never meant to invent a life; it was only meant to express one.

We live in a culture that quietly but relentlessly teaches us we are not enough. Every day, we are surrounded by messages suggesting that if we used the right face cream, our beauty would finally be complete; if we drove the right car, our success would finally be legitimate; if we swallowed the right pill, we would finally become whatever we are currently failing to be. Each of these messages implies the same assumption: as you are now, you are lacking. This narrative finds more than fertile ground, especially because of this unarticulated sense that we're missing something essential. A restlessness. A quiet dissatisfaction. A sadness that seems to come from nowhere. We assume this feeling is personal failure, when in truth it is the very signal of misalignment—the ache that arises when we are not living according to our deeper design.

I know this not as theory but as experience. After surviving profound hardship, I was determined to build a new life for myself. I had never been taught that it was possible to create the life I truly wanted. So, I had never really asked myself what that might even be. I borrowed from what culture had taught me. I envisioned the kind of successful life everybody else wanted. It looked impressive from the outside: wealth, fast cars, and a beautiful home overlooking the ocean. And I achieved it! All of it! Nobody was more surprised than I was! But I also felt strangely hollow. The wild happiness and satisfaction I had imagined never came. I chastised myself for my seeming ingratitude in the face of so much privilege, assuming, as usual, that the constant anxiety, loneliness, and emptiness I felt meant there was something wrong with me. Finally, I came to believe that happiness simply was not possible for me. That's why I thought of suicide.

Conscious Attention

There is another aspect that further exacerbates the terrible conflict within the Four Percent. Quantum physics has shown us that at the level of the quantum field, reality does not exist as fixed outcomes. All possibilities, from the most life-affirming to the most destructive, are present, not as certainties but as probabilities awaiting engagement. That's another way of saying that anything can happen, and potentially does. Each pattern of possibilities has distinct qualities. When attention aligns with a particular frequency, we experience its properties. Fear narrows perception. Coherence expands it. Survival collapses one kind of world. Alignment collapses another. This does not mean we *create reality at the quantum level,* but it does mean that the reality we experience is determined by where we place our focus. Conscious attention collapses the wave.

On the surface, the Four Percent world seems fragmented and competitive. Everything feels separate. I am here. You are there. Resources feel scarce. Safety feels conditional. From this vantage point, attention centers on survival. The nervous system scans for threats, comparisons, and deficiencies. This orientation feels rational because it aligns with what we see and hear every day. But, as real as it may seem, it misunderstands the Four Percent world's purpose. The interface was never meant to be a battleground. It was meant to be a mirror.

When attention is governed by survival, it is drawn to fearful possibilities—not because they are destined to occur, but because they carry great emotional intensity. Repeated focus gives these possibilities weight, and over time, they collapse into lived experience. The world then mirrors the very fears through which it is approached. Reality feels adversarial. Life becomes something to manage rather than inhabit. A subtle but persistent belief takes hold: we may not have what it takes to meet all that lies ahead. This is not because the world is inherently dangerous, but because our attention has become fixed on this narrow spectrum of possibilities.

The Four Percent world is powerful, immersive, and persuasive —but it is not the whole. As real and complete as it seems, it is merely the surface expression of something far deeper. To understand why alignment brings such profound peace and why misalignment produces such persistent dissatisfaction, we must now turn our attention to what lies beneath. That exploration leads us into the Ninety-Six Percent—the deeper dimension of Consciousness this book proposes as the ground from which all expression arises.

Chapter Five

The Fifth Principle

The Quantum Universe is Ninety-Six Percent of Reality

If the Four-Percent world is a mirror, the most important question is what it is meant to reflect. Up to this point, we have examined the interface—how perception is constructed, how thought arises, how our identity becomes entangled with our survival mechanisms, and how suffering emerges when the surface of life is mistaken for its Source. We have looked closely at the machinery of experience: the brain's predictive habits, the nervous system's vigilance, and the mind's tendency to confuse success with truth.

What we have not yet named directly is the dimension beneath the visible that drives it all. In cosmology, although we can't prove or measure the Ninety-Six Percent, we know it is there. Its influence is unmistakably powerful. In human terms, the Ninety-Six Percent is easier to miss. Not because it is abstract or mystical, but because it is so fundamental that it is usually overlooked. Like water to a fish, it is everywhere, rarely questioned, and therefore largely ignored. Most people are unaware of the distinction between the person they think they are and the Presence that gazes out from behind their eyes. We believe ourselves to be alive without realizing that we are life itself.

Looking In All The Wrong Places

In our search to figure out who we are, we naturally turn to what we can see. We look for identity in our personality. We decide who we are based on temperament, emotional patterns, and habitual reactions. If we are confident in certain situations, we conclude we are confident people. If we are anxious, we decide that anxiety defines us. A passing mood becomes a self-concept. "I'm just not assertive." "I've always been sensitive." Personality, which is fluid and context-dependent, hardens into identity.

We search for ourselves in our behavior. We judge who we are by how productive we are, how disciplined we appear, and how well we perform under pressure. A day of procrastination becomes evidence of laziness. A moment of reactivity becomes proof of immaturity. We forget that behavior shifts with fatigue, fear, stress, and circumstance, and instead treat it as a verdict on our worth.

We look for ourselves in history. Past failures quietly dictate what the future can be. Past successes become expectations we feel pressured to meet. Old stories repeat internally—*this is how life goes for me*—and the past begins to colonize the present. We search for meaning in outcomes.

When things go well, life feels meaningful. When they don't, meaning collapses. We learn to equate worth with results, fulfillment with external validation, and coherence with circumstances beyond our control.

We search for identity in roles. Parent. Partner. Professional. Caregiver. Achiever. The role provides structure and social recognition, but over time, it becomes a substitute for presence. When the role shifts or disappears, we feel disoriented, unsure of who we are without it. We also search for purpose in achievement. We assume that if we do enough, reach far enough, or prove ourselves sufficiently, something inside us will finally settle. We chase milestones, believing fulfillment lies just beyond the next accomplishment. When we arrive and still feel unchanged, we assume the problem is that we have not gone far enough yet, or worse, that we have been lied to.

All of these strategies are understandable. They are attempts to locate who we are in what is visible, measurable, and socially reinforced. But there is a common misunderstanding. Personality, behavior, history, roles, success, failure, and circumstances are results, not causes. They are expressions of the surface level of experience. They do not drive things; they are not the underlying origins of life. When we mistake the expression for its Source, there is nothing to hold on to. Identity becomes fragile, and meaning becomes unstable because life's surface is constantly changing.

Expressions of Consciousness

Beneath the interface lies something prior to thought, prior to emotion, and even prior to perception itself. Call it Soul, the Quantum Self, or the Signature Blueprint. The inner architecture is not something we acquire through effort or develop over time. It is what we begin with.

Earlier in this book, we explored a foundational idea: Consciousness is not produced by the brain; it is fundamental. Not emergent, not accidental, not secondary. Consciousness does not arise from matter; matter arises within Consciousness. This position is increasingly recognized as the inevitable conclusion in physics, neuroscience, and philosophy, precisely because material explanations alone have reached their limits. If Consciousness is fundamental—if it is the ground from which experience arises—then it cannot be static. Consciousness is not a thing that simply exists in a fixed state. It is dynamic. It is aware. And like anything aware, it has an intrinsic movement. That movement is toward knowing itself.

This may sound abstract at first, but it is actually quite intuitive. By its very nature, Awareness is awareness *of something*. In our daily lives, Awareness is directed outward —toward objects, thoughts, sensations, and experiences. But if there were nothing external to be aware of, Awareness would, by necessity, turn inward. Consciousness, being all that exists, has nothing external, *no other* to observe. Its only possible object is itself. And so, its primary movement is self-recognition. It is not an intention in the human sense.

Consciousness is not trying to become something it is not. It simply wants to recognize itself in its infinite expressions. The Big Bang is not merely the beginning of matter; it is the first articulation of Consciousness into form. An unfolding. A differentiation. A way for the infinite to experience itself through the finite. Consciousness did not merely create the universe. It became the universe.

It expressed itself in infinite ways, as energy, as matter, and as space-time itself. It became the cosmos—stars, galaxies, planets, minerals. Every physical structure is a configuration of the same underlying intelligence, temporarily taking shape in different arrangements. Life is the next stage of that same expression. Consciousness organizes itself into living systems—plants, animals, ecosystems—each with an increasing capacity to respond, adapt, and experience. Life is not something added to matter. Life is Consciousness expressing itself biologically.

Most likely, there are infinite dimensions; in ours, that expression included us. Consciousness became human not to elevate humanity above all else but to experience itself in a particular way. Through self-reflection. Through imagination. Through language. Through the ability to remember the past, anticipate the future, and wonder about meaning. Through the ability to say, *"I am"* and ask what that means. This does not make humans superior. It makes us specific. We are one expression among infinitely many, each offering a distinct lens through which Consciousness experiences itself.

Everything that exists in the physical world is, at its core, an idea within the *mind of Consciousness*. As mentioned before, an idea is never just a flash of brilliance; it is a completed, organizing principle—a draft that defines its entire integrity. A house cannot be constructed without a fully fleshed-out, detailed blueprint. The blueprint is not the house, but it determines whether the house will stand. If walls are placed where load-bearing structures should be, collapse is inevitable. Not because the house is "bad," but because it violates its design.

The Deeper Design

The same principle applies to living systems. To plants. To animals. And, more importantly, to us. Each of us carries a detailed underlying architecture—a precise internal design that outlines how Consciousness intends to express itself in this particular expression. Naturally, this architecture encompasses our physical characteristics, including sex, height, body type, hair, eye color, and skin color. The design also defines our inner capacities, talents, curiosities, sensitivities, and the things that reliably draw our attention. Like Dark Matter, these characteristics are not truly measurable or visible in the physical world. They can only be recognized by their effects on us. They form our talents, interests, and passions. They determine what feels meaningful and what repeatedly calls to us despite conditioning or expectation. Just as importantly, the architecture also includes our limitations—not as flaws, but as boundaries that give shape to expression. Just as a violin is not meant to function like a drum, we are not meant to excel at everything. Our design is specific, not deficient.

This architecture is not imposed from the outside. It is intrinsic and immutable. It is the source code. Just as our physical characteristics are fixed, so are the structural ones. You can't change the color of your skin and eyes any more than you can change your gifts and talents or the things you care about. You cannot ignore the intrinsic design, at least not successfully. If you turn away, it will always feel as if something is missing. You cannot change what moves you, not authentically. Authenticity is not something we invent through effort or ambition. It is something we discover through attention. When we live in ways that honor the design, life feels congruent and coherent. Energy flows more naturally. Effort feels meaningful rather than depleting. Challenges still exist, but they do not feel like constant resistance. There is a sense—sometimes subtle, sometimes unmistakable—that we are moving in the right direction.

When we live in ways that ignore or contradict this design, suffering arises. This suffering is not punishment. It is not evidence of failure. It is not a moral verdict. It is structural feedback. Just as a building will strain and crack when used contrary to its design, a human life will experience tension, exhaustion, and dissatisfaction when expression is misaligned. What we label as anxiety, depression, burnout, or chronic dissatisfaction is not pathology. It is merely a signal; the system indicates that the surface of life no longer accurately reflects the structure beneath it.

This is where the distinction between the Four Percent and the Ninety-Six Percent becomes rather precise. The Four Percent is the mirror that reflects authentic expression. It is the interface through which life is meant to be enjoyed. The Ninety-Six Percent is the originating principle—the deeper intelligence and potential from which that expression arises. These are not two separate selves. There is no division. There is one whole, one life, one continuity, expressing itself at different levels. The Ninety-Six Percent is not hidden behind the Four Percent. It is expressing itself *through* it. When the expression reflects the origin, we experience wholeness. When it does not, we experience fragmentation.

Alignment, then, is not a spiritual achievement. It is not something we earn, force, or strive for. It is a condition of coherence. It occurs when the Four Percent world is allowed to mirror the deeper architecture it was always meant to express. The Ninety-Six Percent does not need to be activated. It is not in some far-off future. Your Original Blueprint is fully executed and now in effect. It is your highest potential laid out in the finest detail. Here, you are magnificent, as designed by the Consciousness that created the magnificent universe out of itself. When you regard yourself in this light, things shift noticeably.

We often consider ourselves imperfect—unfinished, flawed, in need of correction. This is a grave misunderstanding. At the level of the Ninety-Six Percent—the Signature Blueprint, the originating architecture of our being—nothing is missing, broken, or in need of improvement. Nothing requires control. There is no striving to become more. You are already effortlessly at the height of your being. Here, no one is ever imperfect; all are whole.

What appears as imperfection is visible only at the level of the Four Percent, where expression unfolds over time. The Four Percent world is a domain of becoming. Here, life moves incrementally. Capacity reveals itself gradually. Confidence builds through doing. Understanding deepens through experience. Within this process, we naturally appear unfinished—not because we are flawed, but because we are in motion. Becoming can look like imperfection when mistaken for origin. Would we call a child imperfect simply because it doesn't yet know how to walk? When we judge ourselves solely by what is currently visible—behavior, performance, emotional reactivity, or circumstance—we overlook the deeper structure from which all of that is emerging. We mistake the unfinished sentence for a failed thought. We forget that expression takes time. The Blueprint is already complete; the revelation is evolving.

Sadly, this confusion shapes our self-perception in subtle yet powerful ways. We internalize the idea that we must fix ourselves before we are worthy of expression, contribution, or love. We see growth as repair rather than emergence. We compare ourselves to others, and the comparison can never be fair because it pits our perceived weakness against another's most polished strength. In doing so, we add a skewed perception and an unnecessary struggle to a process that was never meant to feel punitive.

When we dare to see ourselves differently—not as broken beings in need of correction, but as coherent designs in the process of their magnificent articulation—something essential relaxes. Magic happens when we begin to listen within rather than dismiss the still, small voice out of hand. Something long cramped emerges to stretch its wings. Growth no longer feels like a verdict. It feels like unfolding. Effort becomes participation rather than compensation. Self-compassion arises not as indulgence but as accuracy. Our perfection is not static or finished; it is intrinsic. It lives at the design level, not the display level. Recognizing this is not to deny responsibility or effort. It is to remove unnecessary and unreasonable shame from the process of becoming. We are not here to perfect ourselves. We are here to reveal what has always been whole.

Determinism Versus Free Will

The presence of an underlying structure inevitably raises questions about the nature of free will. The Blueprint does not eliminate freedom; it makes freedom possible. Without structure, there is no meaningful choice. A being without form, capacity, inclination, or orientation would not be free—it would be incoherent. Freedom does not arise from randomness. It arises from *Agency within Design*.

We already accept this intuitively in the physical domain. We are born with a specific body: a particular height, a unique nervous system, and a genetic makeup that determines skin color, eye color, and bone structure. We may not always appreciate the cards we were dealt, but we recognize them as the conditions we live with. The body is not a prison; it is an instrument. The same is true at the level of Consciousness.

The Signature Blueprint is not a script that dictates what must happen. It is the underlying architecture that defines *how* Consciousness would express itself through a particular life. It sets parameters, not outcomes. It establishes capacities, not commands. Within those parameters, freedom emerges. You are not free to be anything whatsoever—but you are profoundly free to be *yourself*. That distinction matters. A violin is not free to become a piano, but within its design, there is an extraordinary expressive range. It can play an endless variety of music, interpret the same score in countless ways, or remain silent. The instrument does not negate freedom; it enables it.

Human freedom works the same way. The Blueprint does not force alignment. It does not compel expression. It does not punish deviation. You can ignore, resist, suppress, or distort it—and many people do, though usually not by choice but out of ignorance of its existence. However, that, too, is free will. The experience of that freedom will feel different. It will carry friction, fatigue, and incoherence, not because of moral failure but because structure is strained. Free will, then, is not the freedom to escape Design. It is the freedom to *relate* to Design. We choose whether to follow the Blueprint or override it. Whether consciously or not, we choose whether to act from fear or from coherence. We are free to treat our inner signals as guidance or dismiss them as impractical. We choose how we respond to conditioning, limitations, and circumstances.

At the Four Percent level, free will appears as choice: what we say, what we do, what we pursue, and what we avoid. At the Ninety-Six Percent level, free will appears as orientation: whether Consciousness expresses itself clearly in this life or becomes obscured by misidentification. Free will is not absolute independence from all structure. That would be chaos. Free will is participatory alignment—the capacity to cooperate consciously with the intelligence already at work within us. The deepest loss of freedom does not come from having a Blueprint; it comes from being unaware that one exists.

When we don't know our Design, we confuse conditioning with choice. We mistake reaction for decision. We live as if freedom means doing what we *should,* like eating our vegetables, rather than expressing our greatest fulfillment. The paradox is this: The more clearly we recognize our Blueprint, the freer we become. Not freer to be anything— but freer to be true. Free to be authentic. And that is the only kind of freedom that actually feels like freedom from the inside.

The Greatest Fulfillment

What is often overlooked in discussions of free will is desire itself. Following our Signature Blueprint is not something we are forced to do; it is something we are drawn to. The Blueprint does not impose itself through obligation or command. It expresses itself as attraction, curiosity, and love. It is not experienced as pressure but as preference. We do not all want the same things, and that is precisely the point. What we are most drawn to is not random. It reflects the particular way Consciousness seeks to express itself through each of us. The Blueprint does not tell us what we *should* do. It reveals what we *want* to do—deeply, consistently, and without coercion.

From this perspective, the question of freedom changes. Am I unfree because I am drawn to writing and making art? Or is that attraction itself an expression of my freedom? I am not constrained by my talents; I am animated by them. I do not experience my inclinations as a limitation. I experience them as my greatest aliveness. Freedom is not the absence of preference. It is the ability to honor what moves us without distortion or apology.

We rarely question whether we are *unfree* because we love certain people, places, or ways of being. We recognize love as a choice expressed through attraction rather than force. Our Signature Blueprint works the same way. It is not a rulebook; it is a resonance. It is love itself. This is why alignment does not feel like obedience. It feels like relief. It feels like coming home to a place we've never been before, a place whose call we learned to ignore. The Blueprint is not an external authority; it is the inner logic of what we love most. To follow it is not to surrender freedom; it is to stop resisting it.

The Quantum Self we speak of is not a static template or a rigid destiny. It is better understood as a range of coherent expression—a defined pattern of possibility that seeks to take form through our particular life. This is often sensed as an inner pull. Long before we can articulate it, something in us leans toward certain experiences, skills, environments, and forms of engagement. A child becomes absorbed in music, numbers, movement, building, storytelling, caregiving, or exploration. Before it is taught to lay aside childish things, attention gathers effortlessly. Curiosity sustains itself without reward. Time collapses. These were early signals of design, and the child was so happy.

Alignment does not begin with effort or strategy. It begins with attention. The Blueprint does not announce itself with pressure or command. It speaks quietly through attraction, curiosity, and a subtle yet persistent sense of aliveness. It appears as a small spark of joy—an interest that returns again and again, a pull toward certain activities, ideas, or forms of engagement that feel strangely familiar even when they are new. These signals are easy to dismiss precisely because they are gentle. They do not shout. They invite.

Most of us learn early to override these invitations. We are taught to put ourselves last, distrust our desires, and treat yearning as impractical or irresponsible. We learn to prioritize obligation over inclination and usefulness over love. Over time, the blueprint's signals are not lost—they can never be completely gone—but they become buried. What was once excitement becomes restlessness. What was once joy becomes a vague dissatisfaction we cannot quite name. Alignment begins when we reverse this habit.

It is not about indulging every impulse – "I want a Twinkie, and I want it now!" It is about discerning which impulses are true expressions of our nature. The Soul never pulls us toward what drains us. It draws us toward what we are built to do. A spoon is not tempted to cut, and a knife does not yearn to scoop. Each works best when used as intended. That is literally what they were built for. The same is true for us. What we are meant to do is not arbitrary; it is coherent with how we are made.

When we follow these signals—tentatively at first, then with growing trust—the quality of the effort changes. Work feels absorbing rather than depleting. Learning feels nourishing rather than exhausting. Even discipline changes, because it is no longer imposed from the outside but arises naturally in service of something we deeply care about. This feels like relief. We stop performing a version of ourselves designed to meet expectations and begin expressing something real.

Paradoxically, this makes us more available to others, not less. When we stop suppressing our nature, there is more energy to give. That's why actualized people seem to have boundless energy to get things done. Contribution becomes an extension of coherence rather than an act of sacrifice. Listening to the spark of joy is not a detour from responsibility. It is how responsibility becomes sustainable. The Blueprint does not ask us to abandon the world. It asks us to participate honestly—through the capacities, interests, and forms of expression already alive within us. Alignment, in this sense, is not something we achieve; it is simply what we stop resisting.

Alignment, however, is not a fixed condition. It is dynamic. The Blueprint may remain constant, but our access to it shifts depending on the state we are in. At times, we experience clarity and expansion; at others, contraction and distortion. What changes is not the deeper architecture itself but the frequency with which we perceive and express it.

These States of Consciousness determine which aspects of the infinite field are available to us at any given moment. The next chapter explores these States—not as abstractions but as lived frequencies that shape human experience in immediate, measurable ways. We will examine how they arise, how they influence perception, and how shifting one's state changes how life unfolds. The question is not simply what Consciousness is. The question is: how is it moving through you right now?

Chapter Six

The Sixth Principle

Abundance, Wellness, and Love are States of Consciousness

Dark matter and dark energy—the substance of the Ninety-Six Percent—cannot be directly measured. Science has long been cautious, even dismissive, of phenomena that are intangible or immeasurable. Yet the effects of dark matter and dark energy on the visible universe are so consistent, so powerful, and so unavoidable that their existence cannot be denied. We know they exist because of what they do. Galaxies hold together. Expansion is self-generating and ongoing. Order repeatedly emerges from apparent chaos. Systems self-regulate and balance. Cohesion reasserts itself. Life persists against extraordinary odds because reality itself has dependable default behaviors. Vast, consistent, and beautifully ordered patterns emerge across incomprehensible scales. To call this random would be a disservice to reality.

Whether through telescopes, microscopes, or direct observation of living systems, we encounter one fundamental fact: the universe exhibits inherent characteristics, specific ways in which it reliably behaves. The effects of these behaviors are not merely physical in the conventional sense, like gravity or aerodynamics. They also extend beyond the physical, yielding undeniable, observable results that operate at a deeper level.

Expansion, Balance And Cohesion

Consciousness is energy, and all energy exists at different frequencies. Each frequency carries distinctive information and exhibits definite characteristics, or *States of Consciousness*. There are an infinite number of frequencies, but for the purposes of this book, I will focus on three fundamental universal states that underlie humanity's most persistent concerns.

The rise of the multi-billion-dollar self-help industry reflects humanity's most urgent struggles. Concerns about money, health, and relationships dominate our thoughts, conversations, and choices. Nearly everyone faces financial strain in some form, whether from insufficient income, fear of losing what they have, or exhaustion from work that drains them while offering little fulfillment. An overwhelming share of the population is affected by health challenges, and many people yearn for connection while struggling with loneliness or unfulfilling relationships.

Beneath all this striving lies a single, largely unquestioned assumption: that abundance, wellness, and love are things we must somehow *acquire*. That we must chase them, earn them, or improve ourselves to be worthy of them. And so we run—from book to book, method to method, relationship to relationship—hoping the next strategy will finally deliver what feels missing. But what if nothing was ever missing? What if the reason we yearn so deeply for these qualities is not that they are absent, but that they are ever-present, like water to a fish? What if we are not meant to *create* these states, but simply to *align* with them?

The States of abundance, wellness, and love align directly with the universe's most intrinsic tendencies toward expansion, balance, and coherence. These qualities are intrinsic to the universe and must therefore be present in Consciousness. As reflections of Consciousness, we intrinsically long for these States—not because we lack them, but because they are the very foundation of who we are. We are drawn to them like iron filings to a magnet. The filings respond because the field acts through them. The field reveals itself in the behavior of the material that arises within it. We do not generate them. We do not earn them. They are not conditions we enter and exit. They are inherent, fundamental aspects of who we are.

The Four Percent reality, however, is optimized for survival. It filters experience toward what is immediate, measurable, and threatening. In doing so, it excludes much of what is already present. When awareness is constrained this way, the deeper States of Consciousness recede from view —not because reality has changed, but because perception has contracted. This is not a loss of reality. It is a loss of access. When this occurs, the Four Percent world begins to feel fragmented, effortful, and lacking—not because the universe has withdrawn its inherent qualities, but because attention is no longer oriented toward them.

We can never truly be outside of abundance, wellness, or love, any more than the ocean can be without depth. What *can*—and often does—happen is that our awareness of these States narrows. When awareness narrows, these properties might as well be nonexistent, since they no longer appear in our lived experience. This narrowing of perception is a *defining feature of the Four Percent* world.

When these States fall out of our awareness, we try to manufacture their effects. We work relentlessly to create abundance while feeling internally scarce. We pursue health ceaselessly while living as if breakdown were the natural order. We doggedly seek connection while feeling separate and alone, or worse, unworthy of love. From within this orientation, effort increases—but coherence does not. Outcomes cannot generate the States from which they arise.

To understand why money, health, and relationships feel so difficult at the human level, we must examine the universal States of Consciousness from which these expressions arise.

The State of Abundance

The universe demonstrates extraordinary abundance. The cosmos is so vast, so profuse, so relentlessly generative that language itself strains to describe it. Hundreds of billions of galaxies, each containing hundreds of billions of stars. Energy reorganizes itself continuously. Creation without depletion. Expansion without exhaustion. Space itself unfolds outward, never retreating into conservation. And yet, at the human level, we somehow accept the crazy idea that all this infinite, self-generating, cosmic abundance comes to an unexpected, screeching halt right at our wallets. This contradiction does not exist in reality. It exists only in our perception.

Abundance is not something the universe occasionally produces, nor is it the result of accumulation, effort, or manufacture. Consciousness does not acquire, stockpile, or store. It has no reserves and no scarcity. Abundance is an observable property of Consciousness itself. *Expansion is not an event—it is a condition.* Where infinity exists, limitation cannot be the default state. This governing tendency is visible everywhere. Our solar system spans billions of miles. On Earth, the same principle holds. Life does not merely survive; it proliferates. Species diversify. Ecosystems regenerate. Forests return after fire. Coral reefs rebuild themselves. Seeds push through solid concrete if they must. Even in environments that appear hostile or depleted, life resumes and expands. The dominant movement of nature is not restraint but expression. Abundance, then, is not something Consciousness *produces*. It is something Consciousness *is*. And because we arise from Consciousness, abundance is not something we must obtain. It is intrinsic to us.

Yet this is not how life is typically experienced. In the survival-based Four-Percent world, scarcity often becomes the dominant atmosphere. Time feels scarce. Resources feel fragile. Security feels conditional. Even when there is objectively enough, life is lived as if there is not. This does not mean abundance is ever absent. It indicates that awareness of it has narrowed, and that the distinction is crucial.

The Four Percent mind is simply not designed to perceive abundance. It evolved for survival. Its function is to recognize potential danger, threat, risk, and insufficiency—not to appreciate infinity. This is not a personal failure or a lack of discipline; it is a physiological phenomenon. A nervous system optimized for detecting danger does not automatically register boundlessness. It is built to monitor what might go wrong. When survival becomes the dominant operating mode, perception automatically contracts. Attention narrows. The world appears smaller, harsher, and more limited—not because reality has changed, but because the State of Consciousness through which it is filtered is constrained. Abundance remains fully intact at the deeper level, but it is no longer palpable within that State. This is where confusion arises.

Saying that abundance is a State of Consciousness does *not* mean money appears without action. It means action arises from a different orientation. In a state of abundance, work does not disappear—but it no longer feels like extraction or sacrifice. It feels engaging, generative, and enjoyable. Energy increases with participation rather than being drained by it. Opportunities are noticed, followed, and expanded—not because they magically appear, but because perception is open to them.

By contrast, when someone operates from a sense of scarcity, effort feels heavy and urgent. Work becomes draining, something to escape. The dream is to finally *do nothing, to finally rest, to finally stop.* Ironically, this very longing reveals the absence of abundance. *The system is already depleted.* This is why relentless effort in the face of scarcity must fail. Hard work cannot create alignment. Abundance does not respond to effort. It responds to coherence.

Cultural beliefs such as *"there's never enough" or "money doesn't grow on trees"* do not merely influence thought; they shape perception itself. When these beliefs are active, the experience of lack feels convincing and self-reinforcing. What appears to be an economic problem is, at its core, a matter of perception.

This also explains why movements promising transformation through affirmations and visualization alone—most famously those popularized by *The Secret*—generated enormous enthusiasm, only to be followed by widespread disappointment. The excitement was not naïve. A part of us always remembers its abundant origins and responds instinctively to the promise of alignment. But the disappointment was inevitable because something essential was misunderstood. Everything we experience exists as energy vibrating at a specific frequency.

Abundance is a frequency, but so is scarcity. Beliefs are not merely ideas we hold; they are the frequencies we inhabit. A belief such as "*money is the root of all evil*" does not merely express a moral stance—it places awareness in a frequency fundamentally incompatible with abundance. From within that state, abundance cannot be perceived, accessed, or lived, no matter how sincere or desperate the desire. This is why affirmations usually fail. If one feels compelled to affirm abundance, it is usually because one is aware of its absence. *The affirmation itself arises from scarcity.*

The universe does not respond to words. It does not respond to requests. It does not respond to effort superimposed on contraction. *In fact, the universe does not respond at all!*

It doesn't respond because it doesn't have to. There's nothing for it to do. All frequencies already exist. Everything you could possibly ask for has already been given. Imagine you're standing in front of a giant buffet with every possible delicacy, and you ask the waiter for sushi. The waiter looks at you as if you've lost your mind, then wordlessly points to twenty platters of every variety of sushi you could imagine. In other words, it's all already there; you just have to open your eyes.

The universe operates the same way. It does not reorganize itself to meet our desires. Consciousness isn't a person. Outside of personal experience, it is entirely impersonal. Anything we could possibly want already exists as a potential frequency within Consciousness. Nothing needs to be created or added. No wish or prayer needs to be granted. You always find what you're looking for. The question is: *where are you looking?*

If abundance did not already exist, there would be no way to create it. But if abundance is a natural State of Consciousness, it does not need to be earned. It simply needs to be recognized. Alignment alters perception. Perception alters engagement. Engagement alters experience. We do not experience what we want. We experience the frequency to which we are tuned. Abundance is not something added to life. *It is what life looks like when it is no longer filtered through a lens of separation and survival.*

The State of Wellness

The universe demonstrates an extraordinary capacity for balance, repair, and self-regulation. This is not accidental; it is the very reason the universe persists. Consciousness is infinite. If disorder, imbalance, or dis-ease were intrinsic to it, that flaw would be infinite. Given enough time—indeed, given infinity itself—such a flaw would compound until a total system breakdown. The fact that the universe persists, organizes, and evolves, and has done so for 13.8 billion years, argues strongly against that possibility. Universal Consciousness cannot be fundamentally broken. Its existence depends on it.

Wellness, therefore, is not something Consciousness strives toward. It is a natural state of balance inherent in it. Because we are expressions of Consciousness, well-being is likewise an intrinsic potential frequency within us. This Intelligence is not abstract. It is visible in the body at every moment.

The heart beats rhythmically without instruction. Respiration and temperature adjust continuously to demand. Digestion unfolds with extraordinary precision. Cellular repair, immune response, hormonal regulation, and neural signaling occur in exquisitely coordinated harmony—millions of functions per second across an estimated fifty trillion cells. Even as you read these words, your body corrects imbalances, repairs damage, and maintains stability without conscious direction. This is not a coincidence. It is Intelligence in motion.

When illness arises, attention naturally fixates on the one area of dysfunction. This is understandable. But in doing so, we often overlook the far greater reality: the vast majority of the system continues to function with remarkable accuracy. A few cells may be out of coherence, while trillions remain exquisitely ordered. The exception draws our focus; the harmony is taken for granted.

The body is not powered by willpower. It is powered by Intelligence. That Intelligence is not summoned when we ask for it. It is already flowing—continuously—just as it flows through every living system in the universe. Healing does not occur because Intelligence *hears us and arrives*. It occurs when interference is reduced.

When something feels *wrong* in the body—pain, illness, dysfunction—our instinct is to interpret it as failure or, worse, betrayal. We assume the body has malfunctioned; something has gone wrong that now needs to be corrected, controlled, or addressed. But at the most fundamental level, illness is not a moral verdict, a punishment, or an attack. It is a frequency.

Everything that exists is energy, and all energy exists as frequency. Every frequency carries information. Wellness is one such frequency, and illness is another. When the body expresses illness, it is not rebelling against us; it is faithfully expressing the frequency we are currently aligned with.

Illness is neither *normal* nor inevitable. It is not native. The body is not designed to live outside the frequency of wellness any more than a radio is built to broadcast static. When illness appears, it indicates we have shifted to a frequency other than Wellness—often through Four Percent beliefs rooted in fear, threat, vulnerability, or simply identification with illness.

From this perspective, the way we typically approach healing reveals the problem. We try to *make ourselves well.* We take medication and supplements, exercise, and follow protocols precisely because we believe we are ill. These interventions may manage symptoms and, in many cases, are necessary and lifesaving, but they *do not create* the frequency of wellness. In fact, most of the time, they are pursued

from within the very state that prevents wellness from arising.

This is why no amount of vigilance, supplementation, or
optimization can replace being in the frequency state of
wellness. Health-supportive behaviors may reflect balance,
but they cannot create it. When behavior stems from fear—
particularly fear of death —it reinforces contraction. When
behavior stems from alignment, it effortlessly supports restoration.

The same principle applies to movement. Exercise pursued as punishment strains the body. However, movement
arising from enjoyment and presence strengthens it. The difference is not the activity itself but the State of Consciousness from which it arises.

Pharmacology reflects the same misunderstanding when
it is treated as a solution rather than as support. Medications
can manage symptoms in the Four Percent world, but they
do not establish a Consciousness of wellness. You cannot
generate wellness from within the frequency of illness. This
is the difference between those who recover and those who
do not. It is not willpower. It is not effort. It is not virtue. It
is not luck. It is alignment. Those who recover are not *better*
people; they are people who, consciously or unconsciously,
shifted out of identification with illness and into coherence
with the frequency of wellness. Wellness does not emerge
from forcing the body into compliance. It emerges when
Consciousness realigns.

The work, then, is *not to fight disease* but to recognize and release beliefs incompatible with wellness. These beliefs hold that the body is fragile, that breakdown is inevitable with age, that we are constantly beset by danger on all sides, and that something is fundamentally wrong. Such beliefs constrain Consciousness to a frequency the body cannot transcend.

When a state of internal wellness is reestablished, the body responds intelligently. When awareness aligns with that state, balance reasserts itself. The body does not need instruction; it already knows how to heal. It does not need to be rescued; it needs to be allowed to express the frequency it was designed to inhabit. Wellness is not created; it is restored. Health is not created; it is revealed.

When the body's natural lifespan ends, that transition need not be marked by fear, pain, or struggle. The body—and the mind that arises with it—exist within time. As time-bound structures, they cannot conceive of their own end without fear. They interpret cessation as annihilation. Yet this fear belongs solely to the Four Percent perspective. Consciousness does not share it.

The physical world, what we have called the Four Percent, is only a surface expression after all—a temporary configuration through which Consciousness experiences itself in form. The physical body was never meant to be mistaken for the whole. The Ninety-Six Percent is not bound by time, biology, or physical duration. It is neither born nor does it die. It neither enters nor exits existence. *It simply is.*

When attention is absorbed entirely in the Four Percent world, the illusion of separation feels complete. Identity collapses into the body. Awareness contracts into narrative. Death appears as a devastating loss because form is mistaken for essence. But when focus shifts away from the interface and returns to its Source, this contraction cannot be sustained. Einstein demonstrated that energy cannot be created or destroyed—only transformed. If Consciousness is fundamental and we are expressions of it, then *whatever we are cannot simply vanish*. It must continue, not as form but as a field.

This chapter does not ask the reader to adopt a metaphysical belief. It asks only that we remain consistent. If the Four Percent is an interface rather than an origin, its dissolution cannot be the end of what it reflects. Even as the reflection fades, the source remains untouched. From this perspective, death is not a failure of wellness. It is not the opposite of health. It is the completion of a particular expression within time. And when alignment is present—when identity is no longer confined to the surface—the end of physical form need not be a battle. It can be a release. Not because something is gained, but because nothing essential was ever at risk of being lost.

The State of Love

There is an essential, powerful force in the cosmos that holds everything together. It is cohesive, so systems organize rather than fly apart. It binds, ensuring that life persists and self-repairs rather than collapsing into entropy. It organizes so galaxies can form, atoms remain stable, and ecosystems balance. It even sacrifices, persists, and, at times, overrides self-interest. It generates order from seeming chaos.

On the cosmic scale, we may call it cohesion, attraction, or integration—but this behavior points to a pattern known by another name. Science demands rigor and courage, along with a willingness to accept the answers, regardless of where they may lead. Likewise, rather than skirting issues, intellectual honesty requires that we name this pattern for what it appears to be: love. Love not as metaphor, not as sentiment, and not as a psychological state we invented. Love not merely as a human emotion or social construct but, indeed, as the driving force of the universe that holds everything together.

Whatever we call it for now, the universe behaves in ways that are strikingly consistent with what we recognize, at the human level, as love. Who is to say that the fundamental cosmic frequency of Consciousness is not the same one we experience as love? It acts like love. It organizes like love. It balances like love. It coheres like love. Why avoid the obvious conclusion that the universe behaves as though love were its operating principle, simply because the answer makes us uncomfortable? Because the word love sounds *so terribly unscientific*?

At the human level, love reveals itself with striking clarity. It overrides self-interest. It ignores cost. It defies calculation. A mother does not protect her child because it is efficient or rational; she does so because love reorganizes priorities at a level far deeper than survival. When that bond is severed, the loss can be physically devastating, underscoring that love is not an idea but a force that organizes the nervous system, the body, and the will. Whenever love is present, healing follows. This is not mystical speculation; it is empirically observable. People in stable, loving relationships live longer, recover faster, and regulate stress more effectively. Love strengthens immunity, lowers inflammation, and increases resilience. It acts as a stabilizing field within the body.

Love flows independent of distance or time. When soulmates find one another in ways that feel inevitable, resonant, or *meant to be*, we reach for safer language again: coincidence, chemistry, compatibility, luck. Yet quantum physics already offers a parallel: quantum entanglement, in which particles remain correlated across space and time, responding as one even when separated. We do not dismiss this as fantasy. We accept it as reality. Why, then, do we hesitate to recognize the same pattern in human connection?

Perhaps we resist it because naming love fundamental would require us to acknowledge that the universe is not indifferent and that our deepest experiences may not be incidental. If Consciousness awakened and expressed itself as the universe, what if it did so not out of curiosity alone but out of love? Not sentimentally, but structurally. Not romantically, but coherently. Love is never weak. Love is not optional. Love is not an afterthought. Love is what holds everything together.

At times, it seems events are *guided*. Something appears to *care*. Events may unfold in ways that feel almost magical. Intuition often seems more than a coincidence. When we get out of the way and allow events to take shape rather than forcing them, paths reveal themselves that effort never could. Many people notice, often only in retrospect, that the events that seemed most catastrophic in their lives later proved to be profound reorganizations. Not as consolation prizes or silver linings, but as structural turning points— interruptions that made possible a life that could not otherwise have emerged.

What appeared to be an ending within the Four Percent frame, from a broader perspective, functioned as a reset. This pattern is not uncommon. It repeats across lives and circumstances. People lose jobs that were quietly suffocating them. Relationships collapse that could never have supported growth. Illness forces a reckoning that reorients priorities. Structures fall apart precisely where they can no longer sustain coherence. And again and again, once enough time has passed, people find themselves saying something that feels almost embarrassing to admit, *"It was the best thing that ever happened to me!"*

This understanding is not abstract to me. Homeless and in the throes of drug addiction, I watched my life crash around me. I turned myself in to the authorities, hoping for a fresh start. Instead, I was sentenced to prison. From within the Four Percent frame, it felt like the end. And yet, in hindsight, that rupture was the greatest act of coherence my life had ever known. Prison gave me something I had never had before: time, stillness, and the absence of distraction. It removed me from a trajectory that could not otherwise have been corrected from within. In that enforced pause, I saw my life clearly, dismantled what no longer fit, and began again in deeper alignment. What felt like an ending was, in fact, a reset—one that could not have occurred any other way. I do not say this to romanticize suffering, but to illustrate a pattern I have seen repeatedly since: when a structure can no longer support coherence, it collapses. And when resistance finally gives way, something more truthful can emerge.

System Dependability

Guidance of this kind, however, *does not require intention* in the human sense. It does not require a universe with preferences, plans, or opinions. *Care does not require personality.* The body offers a clear analogy. The body repairs itself, reroutes around injury, isolates infection, restores balance, and regenerates tissue without ever intending to. This does not mean your liver *cares* about you. The immune system does not *deliberate on your progress*. And yet their behavior is undeniably intelligent, protective, and oriented toward restoration. The body behaves as though coherence matters, even though it has no personal awareness of that fact. What if Consciousness behaves in the same way?

In that case, what we experience as *guidance would not be instruction* but simply alignment. Intuition would not be prophecy but access to a broader field of information beyond what linear thinking can process. A happy accident would not be a *miracle,* but a convergence made visible once resistance drops. When effort relaxes and control loosens, the system is free to do what it does best: *reorganize toward coherence.*

This would explain why intuition often appears exactly when needed—not because it is scheduled, but because moments of uncertainty or crisis collapse attention into a single, salient focus. Noise falls away. Awareness widens. Information that was always present becomes accessible. It is simply "what's next" in the program. It would also explain why practices of non-resistance and allowing—found across spiritual traditions—often produce results that force never could. These practices do not summon something new into existence. They remove interference. They stop insisting on a smaller outcome than coherence would naturally produce.

Beauty, too, becomes revealing in this light. Beauty is not required for survival. Symmetry, color, rhythm, and elegance—none are necessary for a functional universe. Yet the cosmos produces them everywhere, lavishly and without restraint. From the structure of galaxies to the patterns of leaves, from the coloration of birds to the mathematics of shells, the universe appears not merely functional but expressive. Expression beyond necessity is not efficiency. It is a delight in articulation.

This suggests that Consciousness is not merely curious but generative in a way that resembles love. Not love as emotion, but love as orientation—the consistent movement toward coherence, continuity, and fuller expression. Seen this way, love is not Consciousness *having feelings*. Love is Consciousness acting in *accordance with its own nature*. When lived from within a human life, that behavior is experienced as care. As meaning. As a sense that life, even when it breaks us open, is not working against us.

This does not mean that everything happens *for* us in a simplistic way. It means that everything participates in a system that does not tolerate incoherence indefinitely. *Structures that cannot sustain alignment eventually dissolve.* Paths that lead away from coherence eventually collapse. And again and again, awareness is invited—gently at first, then with increasing tension—back toward a truer orientation. From within the Four Percent world, collapse looks like failure. In a wider context, it can be an interruption, a redirection, or a release. What feels like punishment may, in fact, be protection. What feels like loss may be the removal of what no longer fits.

If Consciousness is fundamental and expresses itself through coherent laws rather than random chaos, *then the universe does not need to care to behave as if care matters.* Its orientation toward coherence is enough. And when that orientation is lived through a human nervous system, it feels exactly like what we have always called love.

Love, then, is also a State of Consciousness with observable effects. Love is not an outcome; it is a frequency we inhabit. Like all States of Consciousness, it can be experienced only when we are tuned to its frequency. It remains elusive unless we are already in the state of love.

This explains why the longing for love is so intense and persistent. The human desire to love and be loved never fades because it is intrinsic to our nature. We come from it, and when we are no longer oriented toward it, we feel the loss deeply as loneliness. But loneliness is not the absence of another person, even though it feels that way. *Loneliness is the absence of ourselves.* It appears when we can't value ourselves, when we have stopped loving ourselves.

Most people feel they are *not enough.* Scientific studies define this feeling as a deeply ingrained, often learned state of inadequacy linked to trauma, rejection, or chronic self-criticism. They move through life with a quiet assumption that their desires are excessive, their needs unreasonable, and their presence provisional. They don't deserve to ask for what they want. In truth, most do not even know what to ask for, in life or in love—only that they do not want to be alone anymore. I saw this recently when a friend posted on social media, *"Is anyone willing to date me?"* There was no mention of shared values, interests, temperament, or compatibility. Anyone would do. The goal was not connection but relief. And that is precisely the problem. When the primary orientation is toward escaping loneliness rather than meeting another from coherence, the relationship becomes indiscriminate by necessity.

From that State of Consciousness, one can only ever attract someone organized around the same frequency—another seeking rescue from isolation rather than resonance. This is not because lonely people are unworthy of love, but because loneliness itself is a state organized around *absence*. Two people meeting from that state may temporarily ease one another's discomfort, but the underlying loneliness will resurface very quickly. The connection cannot form because it never existed in the first place.

Relationships, like every other outcome, reflect the state from which they arise. When the state is loneliness, loneliness is reinforced. Only when awareness shifts from seeking relief to inhabiting coherence—where one's own presence counts and desire becomes specific rather than desperate—does relationship reorganize into something sustainable.

Now consider what happens when two people meet from within the state of love—not as emotion, but as coherence. In this case, neither person seeks rescue from loneliness nor validation, completion, or being chosen. Each arrives in a state of personal cohesion and balance. Their attention is not organized around urgency but around presence. They are not scanning for approval or gauging how they are perceived. As a result, interaction stabilizes almost immediately. Conversation flows without effort. Silences are not threatening. Differences are delightful, not perceived as dangerous. Signals are readable because they are not distorted by fear or performance. Interest does not spike and collapse; it settles. Attraction does not feel frantic but grounded and unmistakable, at ease. Each person feels more like themselves, not less. Energy is not depleted by the interaction; it is replenished. Time does not feel pressured.

There is no sense of having to secure the connection before it disappears. This is what coherence looks like in a relationship.

Commitment arises naturally, not because it is negotiated or demanded, but because the system supports it. Boundaries clarify rather than harden. Trust develops not as a leap of faith but as a logical response to consistent experience. Love does not feel scarce or fragile; it feels reliable. Importantly, this kind of relationship does not eliminate effort or challenge. Growth continues. Disagreement still occurs, but conflict does not threaten the bond because it was not formed to compensate for a lack. It was formed from symmetry.

This is why relationships formed in the state of love tend to last. Not because the people are better, luckier, or more compatible by chance, but because the underlying frequency supports stability. The outcome always reflects the state. When love is shared as a State of Consciousness, a relationship is no longer something to be secured. It is something that unfolds.

Conclusion

There is no way to be abundant except by being in the state of abundance. Abundance is expressed in its particular frequency, nowhere else. There is no way to be well except by existing in the state of wellness. There is no way to experience love except by being in the state of love. Every other approach is an attempt to reverse cause and effect. We have been taught to believe that if we accumulate enough money, we will feel abundant. We think that if we eliminate symptoms, we will be well. We hope that if we find the right person, we will no longer be alone. But this logic collapses under scrutiny. It assumes that states arise from outcomes, whereas outcomes arise from states.

Trying to produce abundance without being tuned to its frequency is like trying to generate light without electricity. You can polish the bulb, replace the fixture, and rearrange the room—but without current, nothing illuminates. Effort applied without the underlying state cannot succeed, because there is nothing for it to express. This is why so much striving feels exhausting and unrewarding. It is not that people are lazy or misguided. It is that they are attempting to build effects without their underlying cause. Abundance does not appear *after* effort. Effort becomes productive only *from within* the state of abundance.

In the State of Abundance, work does not disappear. It simply changes character. It feels engaging rather than draining. It feels like participation rather than sacrifice. Energy increases with involvement rather than being consumed by it. This is why some people appear to work constantly yet feel energized, while others desperately long for escape. The difference is not the number of hours worked but the state in which the work is performed.

Trying to get rich while living in scarcity is structurally impossible. Scarcity filters perception through a lens of limitation, risk, and insufficiency. From within that state, opportunities are missed, dismissed, or never perceived. The world looks closed because perception is closed.

The same principle applies to health. There is no way to be healthy while worrying about health, whether consciously or unconsciously, and thus inhabiting a state of dis-ease. No amount of supplements, discipline, or vigilance can substitute for the frequency of wellness itself. Health-oriented behaviors can support wellness, but they cannot generate it. Without coherence, the system remains in a compensatory mode—constantly reacting, managing, and bracing.

The same is true of love. *There is no way to experience love while in a state of separation, unworthiness, or fear of connection.* You can date endlessly, optimize profiles, or analyze relationships, but from within those states, intimacy cannot stabilize. The love of your life could be standing right in front of you, and you would not see them – or find them annoying. Love does not arrive and then create openness. Openness allows love to appear.

In every case, trying to create the State through its outcome is an attempt to make something out of nothing. The universe does not work this way. States of Consciousness are primary. Outcomes are secondary. The State is the soil; the outcome is the plant. No amount of pulling on the stem can replace fertile ground.

This is why affirmations fail, why effort exhausts, and why discipline alone collapses. The State has not changed, so the results cannot either. To inhabit abundance is not to stop working. It is to work *from* abundance. To inhabit wellness is not to ignore the body. It is to allow the body's inherent Intelligence to express itself. To inhabit love is not to wait passively but to recognize the beloved. Nothing is created from outside the State that gives rise to it. It is how coherent systems operate.

Chapter Seven

The Seventh Principle

Portals Exist That Connect the Four Percent and the Ninety-Six Percent

Joseph Campbell urged us to *"Follow our Bliss,"* and no words were more important. He was not advocating indulgence. He was pointing to a feedback mechanism that indicates when a life is being lived in accordance with its deeper design. Bliss is not a goal, a reward, or a personality trait; it is a signal. It is the subjective experience of systemic alignment—what it feels like when the visible aspects of life are organized according to the underlying Blueprint. Bliss follows alignment; it does not cause it. When we are in alignment with our original idea, there is a sense of fulfillment and joy. So, following our bliss is the quickest way to learn who we are.

Most people move through life unaware that they possess an underlying architecture. Before encountering a framework like the Iceberg Principles, people often dismiss their preferences as trivial, their desires as unreliable, and their longings as impractical or self-indulgent. They may have a vague sense that certain activities energize them while others exhaust them, and that certain environments feel right while others feel constricting, but they lack a coherent explanation for why or what it means. As a result, powerful signals are often overridden by obligation, efficiency, or external approval. It is not that people are out of touch with themselves. The problem is that they have never been taught to interpret the signals they continually receive. The underlying Blueprint is not physical; therefore, its signals cannot be accessed directly. Like the unseen forces that hold the cosmos together, they can only be inferred through their effects.

There are, however, conditions under which these influences become perceptible, portals where the boundary between the visible Four Percent and the invisible Ninety-Six Percent thins. Physics itself hints at such thresholds: black holes suggest vast regions beyond our understanding. They expose the edges of physical reality, where information, not matter, appears to be primary. Some portals are literal and physical. Sexual union, for example, is the only possible gateway through which Consciousness can enter embodied form. Some portals are purely energetic. Meditation clears static interference, gratitude and appreciation elevate perception, and intuition is the pathway. Each portal offers powerful inroads into the deeper organization beneath the surface.

The Iceberg Principles is not a self-help book. For anyone interested in learning more about ways to connect the

Four Percent and the Ninety-Six Percent, further study is needed. You may find more links, tips, and information on my website, MarniSpencerDevlin.com.

Sex — Energetic Propulsion

The first portal is the most fundamental, as it is the only avenue through which Consciousness enters from the Ninety-six Percent into the Four Percent. Creation itself passes through this threshold. Therefore, sex is not merely a biological function or a recreational act; it is one of the most powerful energetic portals in the universe. Sexual energy is the generative force that brings Consciousness into form, and when engaged consciously, it can temporarily lift the individual beyond ordinary, time-bound perception into an expanded state of being.

Sex is the glue that holds a healthy relationship together. Scientific evidence shows that regular sexual activity offers significant physical and mental health benefits, including reduced stress, boosted immunity, improved heart health, and stronger bonding. Sex triggers the release of endorphins and hormones such as oxytocin and dopamine, which lower cortisol, alleviate pain, improve sleep, and enhance mood.

Yet for most people, this vital portal remains forever closed. Conditioned by centuries of religious dogma, moral restrictions, and cultural shame, sex has been reduced to a brief discharge of tension—a mere way to quiet stress rather than the Consciousness-expanding energy it is. When sexual energy is condemned or feared, it is trapped in the lower layers of the Iceberg, where it tends to manifest as compulsion, shame, fixation, or dissociation.

In this way, repression and excess are not opposites but mirrors: both reflect sexual energy held below awareness, unable to complete its natural ascent into coherence. When an entire culture frames powerful sexual energy as dirty, dangerous, or morally suspect, people either repress it or compulsively fixate on it, never pausing to appreciate its power. As a result, few ever learn the art of sex as an energetic state. They remain unaware that sexual arousal, when sustained and refined rather than rushed toward release, can open a doorway to heightened awareness, profound intimacy, and states of Consciousness far beyond ordinary pleasure.

In certain streams of Hinduism and Buddhism, sexual expression was neither dismissed nor denied but approached with reverence and discipline. Tantric traditions, in particular, recognized sexual energy as a form of spiritual alchemy —a raw, life-giving force capable of being transmuted rather than expelled. The aim was neither indulgence nor abstinence but transformation. Rather than pursuing momentary orgasmic relief, practitioners sought a sustained state of bliss in which the boundaries of the individual self temporarily dissolved.

In this state, sex becomes less about performance or gratification and more about Presence. The body is no longer treated as an object but as a temple; union is no longer mechanical but ritualistic. Attention slows. Awareness deepens. Energy circulates rather than collapses. When this happens, the ego loosens its grip, and the familiar sense of separation begins to soften. For fleeting moments, identity expands beyond the personal story into something vast, intimate, and luminous. These moments—transitory as they may be—offer a direct glimpse of the expanded state outlined in the Blueprint.

Seen this way, sex is a threshold that can briefly return us to the same generative field from which we emerged. In Zen Buddhism, an orgasm is understood as a brief moment of satori, that is, oneness with the Divine. Because it originates below thought, it can briefly lift Consciousness above thought. In Iceberg terms, it creates a vertical surge—an internal ascent from instinctual depth toward expanded Awareness. This is why sexual states, even in ordinary contexts, are often accompanied by a temporary loss of time, self-consciousness, and narrative identity. The ego loosens. The sense of separation thins. The individual leaves the habitual Four-Percent framework and enters something more expansive. Although the experience is brief, the portal and its Presence remain.

Meditation — Energetic Clearance

Meditation is perhaps the best-known and most intentional portal into the Ninety-six Percent because it does not stimulate energy—it *removes interference*. It works by subtraction. It quiets the brain's relentless self-referential activity, creating the conditions for the deeper structure to become perceptible.

The Iceberg Principles point to a fundamental fact: there are two aspects to the human experience—the physical self, shaped by sensation, memory, and identity, and the quiet Presence of Consciousness within, the *I Am*. Under ordinary circumstances, the surface mind dominates our awareness. It constantly predicts, evaluates, narrates, and rehearses identity, filling Consciousness with cognitive noise. This activity is not inherently problematic, but its constant nature effectively drowns out the Blueprint's subtler signals that arise beneath the surface.

Meditation does not force the mind to be silent; instead, it gently shifts attention. By directing focus toward the *I Am* Presence, thoughts, emotions, and images produced by the mind are recognized as the passing illusions they truly are. They are no longer inhabited but observed, passing through Awareness like clouds through an open sky. As attention disengages from identity maintenance, the Four Percent relaxes its grip. In this quieter internal landscape, signals from the Ninety-six Percent—bodily intelligence, emotional information, intuitive knowing, and creative insight—begin to rise into Awareness.

What emerges is not emptiness but coherence. The signal-to-noise ratio shifts, and the deeper system becomes audible. In this space, we reach the highest expression of our being, yet there is no effort at all. Here, there is no striving to be better, no need for improvement or control. Here, we are *being without becoming*. Meditation thus functions not as an escape from the self but as a reorientation toward its true Source—a deliberate return to the internal architecture from which clarity, alignment, and recognition naturally arise.

Gratitude and Appreciation — Energetic Elevation

Gratitude is a powerful, energetic state that fundamentally alters how we perceive and participate in reality. Research from the HeartMath Institute shows that when we experience gratitude, the heart and brain synchronize, producing energetic coherence that supports healing, clarity, and creative perception. Neurologically and chemically, gratitude signals safety, wholeness, and abundance, and it restores nervous system regulation. But its significance extends beyond biology. As a portal, gratitude aligns us more closely with the expanded mode of being outlined in the Blueprint. It places us in resonance with a broader field of possibility, where intuition sharpens, and perception opens.

Cultivated daily, not as a technique but as an orientation, gratitude becomes a reliable gateway—one that lifts us into a more expansive relationship with ourselves, with life, and with the unseen pattern seeking expression through us. Gratitude isn't manners, mood, or morality—it's *architecture*. A doorway and a state that reorganizes perception and energy at once.

Gratitude is a powerful energy, but there is an even more formidable energy, and that is appreciation. The difference is subtle yet potent. Gratitude tends to be a feeling summoned in response to favorable circumstances. It is about being thankful *to* someone or *for* something. This makes it transactional, relational, or situational. Therefore, it can fall short in its usefulness when you cannot conjure the feeling, which is when you most need this portal.

Appreciation is a coherent internal state that can be entered deliberately and should be cultivated daily as a matter of course. In real estate, appreciation simply means an increase in value. Conversely, when you appreciate what you have, it becomes worth more. When you appreciate your work, it becomes more enjoyable. When you appreciate someone in your life, they become more attractive. Appreciation means looking for the most positive aspects in any given situation, and that requires nothing more than your intention to do so. When sustained, this state measurably elevates our energy, broadens the horizons of perception, and moves us out of contraction and survival-based awareness.

Intuition

Among all the portals, intuition is the most immediate and the most overlooked. It is the language of the Blueprint, not as external guidance but as the natural evolution toward the next step. Intuition does not require extraordinary routines or rituals. It is always there, operating quietly beneath the surface of daily life. We're just not used to listening for it. We mistrust or dismiss the inner voice.

My recognition of this did not begin as a philosophical insight. It began as a practical necessity. After being told I might have only a year to live, I resolved to spend whatever time remained doing only what brought me joy. What I discovered almost immediately was that I did not know how to answer that question. I had spent a lifetime responding to external demands and internalized expectations. My own preferences had been ignored for so long that they no longer registered clearly. I did not yet understand that fulfillment required alignment with a deeper architecture.

I began with the smallest possible experiments. I asked myself simple questions: when I got up in the morning, whether I wanted a cup of coffee or to brush my teeth first. Then, did I want to meditate or go for a walk instead? I have come to believe it is essential to ask such simple questions, at least until we become more adept at interpreting intuitive signals. I continued throughout the day, asking myself these seemingly insignificant questions and deciding what felt right for me from moment to moment. To my surprise, I always recognized a clear preference. There was a clear sense of rightness in one direction and subtle resistance in the other. As I continued to hone my intuition in this way, something unexpected happened. I began to feel more present and noticeably calmer. The internal critical voice quieted. Anxiety diminished. Even the loneliness that had always been my default state vanished. It seemed that by honoring my own voice, even in this simplest of ways, I was becoming a friend to myself. I was not yet following bliss in any dramatic sense, but I was learning to recognize coherence.

One of the most common obstacles to listening is the belief that signals must be justified. Many people wait to honor an inner response until they can explain it logically or defend it socially. This habit effectively silences the Blueprint. The deeper organization of the self does not argue its case. It simply registers. Learning to listen means allowing signals to exist before justifying them.

It is also essential to understand that listening does not automatically require withdrawal from life. We don't all have to become monks or Mother Teresas. Fulfillment is always the result of radical authenticity. Through repeated exposure to one's inner voice, individuals can learn to distinguish between alignment and misalignment, effort and resonance, and who they have been trained to be and who they are designed to express.

When the Blueprint is honored, bliss follows naturally—not as excitement but as deep joy and passion. It is the sense that effort and direction are aligned and that energy flows rather than is forced. Bliss does not tell you what to do next. It simply indicates that the system is coherent. At that point, nothing more needs to be added.

Each of these states offers a brief window in which the Blueprint can be sensed as orientation toward or away from coherence. Of course, many other portals exist, perhaps as many as there are expressions of Consciousness. We are each entirely unique and may find unique ways to access our deeper state. One thing is true for us all: expansion, ease, vitality, and resonance indicate alignment, while contraction, heaviness, and resistance indicate misalignment. The right direction energizes; the wrong way deadens us. These signals are subtle and easily overridden by habit, yet they are consistent for all of us.

The Iceberg Principles do not ask you to become someone new. They ask you to recognize the structure that has been there all along—and to learn to listen when it speaks.

And then, quietly, to live accordingly.

References

Cosmology & Dark Matter / Dark Energy

Planck Collaboration. (2018). Planck 2018 results. VI. Cosmological parameters. Astronomy & Astrophysics, 641, A6.

Riess, A. G., et al. (1998). Observational evidence from supernovae for an accelerating universe and a cosmological constant. The Astronomical Journal, 116(3), 1009–1038.

Perlmutter, S., et al. (1999). Measurements of Ω and Λ from 42 high-redshift supernovae. The Astrophysical Journal, 517(2), 565–586.

Rubin, V. C., Ford, W. K., Jr. (1970). Rotation of the Andromeda Nebula from a spectroscopic survey of emission regions. The Astrophysical Journal, 159, 379–403.

Zwicky, F. (1933). Die Rotverschiebung von extragalaktischen Nebeln. Helvetica Physica Acta, 6, 110–127.

Quantum Mechanics, Measurement, and Decoherence

Zeh, H. D. (1970). On the interpretation of measurement in quantum theory. Foundations of Physics, 1, 69–76.

Zurek, W. H. (1981). Pointer basis of quantum apparatus: Into what mixture does the wave packet collapse? Physical Review D, 24(6), 1516–1525.

Zurek, W. H. (1982). Environment-induced superselection rules. Physical Review D, 26(8), 1862–1880.

Zurek, W. H. (2003). Decoherence, einselection, and the quantum origins of the classical. Reviews of Modern Physics, 75(3), 715–775.

Heisenberg, W. (1958). Physics and Philosophy: The Revolution in Modern Science. Harper.

Bohr, N. (1935). Can the quantum-mechanical description of physical reality be considered complete? Physical Review, 48, 696–702.

Neuroscience: Predictive Processing, Attention, and Emotion

Helmholtz, H. von. (1867). Handbuch der physiologischen Optik (Treatise on Physiological Optics).

Rao, R. P. N., & Ballard, D. H. (1999). Predictive coding in the visual cortex: A functional interpretation of some extra-classical receptive-field effects. Nature Neuroscience, 2, 79–87.

Friston, K. (2005). A theory of cortical responses. Philosophical Transactions of the Royal Society B, 360(1456), 815–836.

Friston, K. (2010). The free-energy principle: A unified brain theory? Nature Reviews Neuroscience, 11, 127–138.

Seth, A. (2021). Being You: A New Science of Consciousness. Faber & Faber.

Moruzzi, G., & Magoun, H. W. (1949). Brain stem reticular formation and activation of the EEG. Electroencephalography and Clinical Neurophysiology, 1, 455–473.

LeDoux, J. E. (1996). The Emotional Brain: The Mysterious Underpinnings of Emotional Life. Simon & Schuster.

Perception: Vision, Color, Sound

Findlay, J. M., & Gilchrist, I. D. (2003). Active Vision: The Psychology of Looking and Seeing. Oxford University Press.

Eagleman, D. (2011). Incognito: The Secret Lives of the Brain. Pantheon.

Land, E. H. (1959). Experiments in color vision. Scientific American, 200(5), 84–99.

Newton, I. (1704). Opticks. (Original publication).

Rayleigh, J. W. S. (1877). The Theory of Sound. Macmillan.

Philosophy of Mind: Qualia and the Hard Problem

Lewis, C. I. (1929). Mind and the World-Order: Outline of a Theory of Knowledge. Scribner.

Nagel, T. (1974). What is it like to be a bat? The Philosophical Review, 83(4), 435–450.

Chalmers, D. J. (1995). Facing up to the problem of consciousness. Journal of Consciousness Studies, 2(3), 200–219.

www.ingramcontent.com/pod-product-compliance
Lightning Source LLC
Chambersburg PA
CBHW060147200526
45165CB00023B/963